This book is to be returned on or before

Surveys in Industrial Wastewater Treatment
Vol. 2 Petroleum and Organic Chemicals Industries

VOL 2

Surveys in Industrial
Wastewater Treatment

Petroleum and Organic Chemicals Industries

Edited by
D Barnes
University of New South Wales

C F Forster
University of Birmingham

S E Hrudey
University of Alberta

π

Pitman Advanced Publishing Program
Boston · London · Melbourne

PITMAN PUBLISHING LIMITED
128 Long Acre, London WC2E 9AN

PITMAN PUBLISHING INC
1020 Plain Street, Marshfield, Massachusetts 02050

Associated Companies
Pitman Publishing Pty Ltd, Melbourne
Pitman Publishing New Zealand Ltd, Wellington
Copp Clark Pitman, Toronto

First published 1984
© Pitman Publishing Limited 1984

Library of Congress Cataloging in Publication Data
Main entry under title:
Petroleum and organic chemicals industries.
  (Surveys in industrial wastewater treatment; vol. 2)
  Includes bibliographical references and index.
  1. Petroleum chemicals industry—Waste disposal.
  2. Dye industry—Waste disposal. 3. Synthetic fuel
  industry—Waste disposal. 4. Sewage disposal.
  I. Barnes, D. II. Forster, C. F. (Christopher F.)
  III. Hrudey, S. E. (Steve E.) IV. Series.
  TD899.P4P48 1984   628.1'6836   84-3163
  ISBN 0-273-08588-3

British Library Cataloguing in Publication Data
Surveys in industrial wastewater treatment.
  Vol 2. Petroleum and organic chemicals
  industries
  1. Water—Purification
  I. Barnes, D.   II. Forster, C. F.
  III. Hrudey, S. E.
  628.1'62   TD430

  ISBN 0-273-08588-3

All rights reserved. No part of this publication may be reproduced, stored in a retrieval system, or transmitted, in any form or by any means, electronic, mechanical, photocopying, recording and/or otherwise, without the prior written permission of the publishers. This book may not be lent, resold or hired out or otherwise disposed of by way of trade in any form of binding or cover other than that in which it is published, without the prior consent of the publishers.

Printed in Great Britain at The Pitman Press, Bath

Preface to the series

THE CONCEPT of a series of books on aspects of industrial wastewater treatment was developed during 1979. At that time one of the editors (DB) was on study leave at the University of Birmingham and another (SEH) had occasion to make several visits to Britain. All of the editors had some experience of treating industrial wastewaters and all had experienced difficulties in obtaining detailed information both about wastewater characteristics and about the performance of particular processes.

The series reviews in detail the treatment of wastewaters from groups of industries. Authors have been selected who can provide a balanced summary based upon their direct experience. Each has been asked to discuss a subject in sufficient depth that readers can obtain a reliable insight into the options available. As such, the individual reviews are not summaries of specific research endeavour: rather they reflect established or proven practice. Equally, the reader has been assumed to have a basic understanding of wastewater treatment, such that the principles of analytical procedures and unit operations need not be explained, except as they have specific features unique to a given industrial wastewater.

In order to provide some cohesion to each volume, related subjects have been grouped together. The first volume is concerned with the food and beverage industries, with chapters on sugar, dairy, beverage, fruit and vegetable and meat and poultry industry wastes. The second volume groups together organic-based wastewaters, with chapters on the dyestuffs, petrochemicals, oil refining and synthetic fuels industries. The third volume, which deals with inorganic wastewaters, covers plating, silver recovery (particularly from the photographic industry), general inorganic chemical industries, chloralkali (particularly the treatment and disposal of mercury sludges) and the steel industry. Subsequent volumes will group topics in a similar manner.

The editors would like to express their thanks to the authors for providing the manuscripts and for tolerating the corrections, threats and browbeatings as editorial deadlines approached and passed. Also we thank the publisher,

Pitman, for help and cooperation and our students and professional colleagues who provided the necessary stimulus for such an endeavour. Finally we express our thanks to our wives and families for their help and support.

DB *Sydney*
CFF *Birmingham*
SEH *Edmonton*

Preface to Volume 2

A BOOK on industrial treatment devoted to the organic chemical industries provides some interesting and rewarding challenges. The past decade has seen remarkable advances in various fields of technology. Many of these have contributed directly to an increased awareness of, concern about and response to the problems and hazards with which wastewater confronts the organic chemical industries.

Firstly, there has been a notable series of advances in analytical technology for organic compounds in waters and wastewaters. Following the widespread adoption of the total organic carbon analyzer in the early 1970s, analytical methods for specific organic compounds have been developing at a rapid pace. The major advances are mainly due to the common use of gas chromatography (GC) in water and wastewater analysis. This technology has more recently been enhanced by the accessibility of capillary column GC which provides greatly enhanced resolution for complex wastewaters. High performance liquid chromatography (HPLC) has made significant advances. It may be hoped that future improvements in detector sensitivity will allow this technology to play an expanded role in treatment evaluation. Finally, the widespread availability of gas chromatography/mass spectrometry (GC/MS) has opened new horizons which were beyond the scope of most wastewater specialists a decade ago.

However, at the same time GC/MS has opened our eyes to a Pandora's box of toxic organic compounds in industrial wastewaters and receiving waters. The finding of trace levels of countless organic compounds in water supplies and sources has led to expanded efforts to assess their individual significance to health. Specifically, our ability to assess the genotoxicity of organic compounds has markedly increased. Long-term concerns with cancer, birth defects and mutations have been addressed with the development of short-term cellular level test procedures to screen organic contaminants. Such techniques are providing a basis for developing treatment technology to deal with the potentially most dangerous compounds.

As our awareness of the scope of organic chemical treatment needs has developed, we have seen major advances in the capabilities of wastewater treatment technology. We have developed a better understanding of the fundamentals of biological treatment, both aerobic and anaerobic. This knowledge has given rise to an expanded array of new treatment options ranging from a variety of fixed biofilm processes to more efficient oxygen and aeration systems. Furthermore, specific organic removal processes such as activated carbon and polymeric resin adsorption have come of age and are being widely adopted. There is also rapid development on other frontiers such as membrane processes. Overall, the future holds tremendous potential for cost-effective, reliable technologies to remove toxic organic compounds from industrial wastewaters.

Finally, we have seen great advances in the industrial process technologies themselves. General consumer demand and newly emerging industries such as microelectronics have introduced new products requiring new processes which, in turn, present new treatment challenges. At the same time, rapid increases in energy costs coupled with dwindling water resources have encouraged a significant trend towards industrial water conservation and reuse. Such reuse options have often proven more challenging than conventional discharge options and improved treatment techniques have emerged.

The rapid pace of development in this field means that it is becoming increasingly difficult for the student to assimilate or the professional to sustain relevant knowledge. In response, this volume attempts to address recent developments and provide the professional practitioner and the specialist student with a statement of current status and an indication of future trends.

Four major organic chemical industries are addressed: petroleum refining, petrochemicals, dyestuffs and synthetic fuels. The last, in particular, comprises emerging technologies that are likely to form a major portion of the industrial landscape in the future. With continued commitment from qualified specialists, we can hope to rise to meet these future challenges.

DB *Sydney*
CFF *Birmingham*
SEH *Edmonton*

Contents

Preface to the series   v

Preface to Volume 2   vii

Abbreviations   xiv

Introduction   xv
D Barnes, C F Forster and S E Hrudey

1 **The management of wastewater from the petroleum refining industry**   1
   A S Vernick, B S Langer, P D Lanik and S E Hrudey

   1.1 Industry background   1
   - 1.1.1 Production processes   1
   - 1.1.2 The products   13
   - 1.1.3 Profile of the industry   14

   1.2 Wastewater characteristics   15
   - 1.2.1 Sources of process wastewater   16
   - 1.2.2 Quality of process wastewater   17
   - 1.2.3 Quantity of process wastewater   18
   - 1.2.4 Quality of treated effluent   21

   1.3 Discharge standards   25
   - 1.3.1 Introduction   25
   - 1.3.2 Petroleum refinery discharge standards   26

   1.4 Control and treatment technology   29
   - 1.4.1 In-plant control   29
   - 1.4.2 In-plant treatment   34
   - 1.4.3 End-of-pipe control technology   37

1.4.4 Summary of effluent treatment methods 48
1.4.5 Residuals management 50

1.5 Costs 52
    1.5.1 Recycle and reuse 53
    1.5.2 In-plant control 53
    1.5.3 End-of-pipe treatment 53

1.6 Summary 57

References 63

2 The treatment of wastes from the synthetic fuels industry 65
R D Neufeld

2.1 Introduction 65

2.2 Coal conversion processes 66
    2.2.1 General considerations 66
    2.2.2 The integrated coal refinery 67
    2.2.3 Coal gasification 67

2.3 Coal conversion plant—sources of wastewater 70
    2.3.1 The overall operation 70
    2.3.2 Coal gasification facilities 74
    2.3.3 Coal liquefaction facilities 77

2.4 Characteristics of coal conversion effluents 79
    2.4.1 Composition by concentration 87
    2.4.2 Waste streams generated by synthesis operations in indirect liquefaction facilities 89
    2.4.3 Specific constituent analysis 90

2.5 Effluent standards 99
    2.5.1 US guidelines for comparable industries 99
    2.5.2 Environmental impact—data and technological development 101

2.6 Design of coal conversion wastewater treatment systems 102
    2.6.1 Principles and treatment methods 102
    2.6.2 Treatment train design: examples 106
    2.6.3 Design of biological reactors 110
    2.6.4 Pilot-plant bioreactor data—process stability and performance 112

2.7 Wastewater treatment costs for coal conversion facilities 116
    2.7.1 Basis of cost assessment 116
    2.7.2 Gravity separation techniques 117

2.7.3 Steam stripping 117
2.7.4 Solvent extraction 117
2.7.5 Biological treatment 117
2.7.6 Carbon adsorption 118
2.7.7 Wastewater treatment train 118

2.8 Solid waste generation, treatment and disposal 118
2.8.1 Sources of coal conversion residuals 118
2.8.2 Gasifier ash or slag 120
2.8.3 Wastewater treatment plant sludges 121
2.8.4 Control options for solid waste disposal and leachate treatment 123

2.9 Summary 125

References 126

3 The treatment of wastes from the petrochemical industry 130
G E Chivers

3.1 The petrochemical industry 130
3.1.1 Introduction 130
3.1.2 Petrochemicals and petrochemical processes 131
3.1.3 Wastewaters from petrochemical works 134

3.2 Environmental pollution from petrochemical works wastewaters 135
3.2.1 Oil pollution 135
3.2.2 Chemical pollution 139
3.2.3 Physical forms of pollution 143

3.3 Legal constraints on wastewater discharges 144
3.3.1 European regional legislation 144
3.3.2 United Kingdom legislation 145

3.4 Standards for discharge 146
3.4.1 General principles 146
3.4.2 Wastewater discharge standards for oil 146
3.4.3 Standards for other constituents 148

3.5 The treatment of petrochemical works wastewaters 152
3.5.1 Treatment principles 152
3.5.2 Wastewater monitoring 154

3.6 Physicochemical treatment processes 155
3.6.1 Steam distillation 155
3.6.2 Treatment to remove less volatile constituents 156
3.6.3 Treatment to remove oil 160

xii CONTENTS

3.7 Biological treatment processes 167
  3.7.1 Pretreatment 167
  3.7.2 Biological treatment of low- and moderate-strength wastewaters 168
  3.7.3 Biological treatment of high-strength wastewaters 172
  3.7.4 Final effluent treatment 181

3.8 Disposal of concentrated wastes and sludges 182

3.9 Costs of effluent treatment 184

3.10 Case studies of wastewater treatment 186
  3.10.1 A resins manufacturing plant 186
  3.10.2 A petrochemical complex 188

3.11 Conclusion 189

Acknowledgements 189

References 189

4 **The treatment of dyewastes** 191
B D Waters

4.1 Introduction 191

4.2 Effluent standards 191

4.3 Problems posed by dyewastes 194
  4.3.1 Impact on receiving water 194
  4.3.2 Colour measurement 196

4.4 The textile industry 196
  4.4.1 Preliminary textile-finishing processes 198
  4.4.2 Constitution of dyewastes 199

4.5 Treatment of dyewastes 200
  4.5.1 Preliminary considerations 200
  4.5.2 Conventional treatment processes 201
  4.5.3 Polishing treatment processes prior to discharge 207
  4.5.4 Adsorption processes 208
  4.5.5 Chemical oxidation 212
  4.5.6 Other treatment processes 214
  4.5.7 Sludge disposal 215

4.6 Complete treatment plant design   216
    4.6.1 General principles   216
    4.6.2 Discharge to sewer   218
    4.6.3 Discharge to a watercourse   218
    4.6.4 Recovery of water   219

4.7 Summary of treatment processes   220

4.8 Case studies   220

4.9 Future developments and trends   227

References   228

Abbreviations

ADMI	American Dye Manufacturers Institute
APHA	American Public Health Association
API	American Petroleum Institute
ASTM	American Society for Testing and Materials
BASF	Badischer Anilin und Soda Fabrik
BAT	'Best available technology'
BOD	Biochemical oxygen demand
BOD_5	5-day biochemical oxygen demand
BPT	'Best practical treatment'
COD	Chemical oxygen demand
DAF	Dissolved air flotation
EDS	Exxon Donor Solvent
EEC	European Economic Community
EPA–'EP'	US EPA's 'Extraction Procedure' to generate leachate from solid waste
FGD	Flue gas desulphurization
GFETC	Grand Forks Energy Technology Center (North Dakota)
ICI	Imperial Chemical Industries
IGT	Institute for Gas Technology (Chicago)
LC_{50}	Lethal concentraton to 50% of exposed organisms
MAF	Moisture- and ash-free
MEA	Monoethanolamine
METC	Morgantown Energy Technology Center
MLSS	Mixed liquor suspended solids
OECD	Organization for Economic Cooperation and Development
PACT	Proprietary powdered activated carbon activated sludge system
RCRA	Resource Conservation and Recovery Act (USA)
ROM	Run-of-mine
SFBG	Stirred fixed bed gasifier
SRC	Solvent refined coal
TDS	Total dissolved solids
TOC	Total organic carbon
TOD	Total oxygen demand
TSS	Total suspended solids
US DOE	United States Department of Energy
US EPA	United States Environmental Protection Agency

Introduction
Principles of industrial wastewater treatment

D Barnes, C F Forster and S E Hrudey

THE WIDE range of industrial manufacturing processes generates an equally diverse range of wastewaters. Clearly it is inappropriate to assume that similar wastewaters will be generated in, say, a brewery and in the manufacture of steel, or that the wastewaters can be treated by the same unit processes designed to the same criteria. However, some aspects of both waste generation and treatment are common to many industries. This brief introduction attempts to summarize some of these common aspects.

Wastewater characteristics

Industrial wastewaters tend to be characterized by great variability in both flow and composition. Only a small number of industrial plants operate continuously to generate wastewater of non-varying characteristics. It is only very large organizations, such as oil refineries, that attempt to maintain a continuous throughput of consistent quality and so can produce a consistent effluent. The majority of industries are small-to-medium sized, do not operate 24 hours a day and do not attempt to produce either product or effluent continuously or consistently.

Many dairies, for example, work for only 8–12 hours a day so only generate wastewater during that period. Several of the effluent-producing operations, such as tank cleaning and disinfection, are discontinuous and hence give rise to variability of both flow and load for treatment. Often the dairy manufactures several products, such as a range of conventional milks (homogenized, high or low butterfat), condensed or evaporated milk, cream products and more specialized products such as yoghurt and cheese. Each of these manufacturing processes produces wastewaters of different volume and composition. Thus, while the BOD_5 received for treatment may average 3000 mg l^{-1} over a working day, this may include periods when the BOD_5 exceeds 20 000 mg l^{-1} or contains high concentrations of sodium hydroxide with associated high pH values.

This variability of industrial wastewaters is accentuated by the short sewer lengths through which wastewater is conveyed to the treatment plant. Municipal plants receive wastewater that has been subjected to considerable mixing, dilution and attenuation in the sewerage system, whereas industrial plants are served by sewers that rarely exceed 1 km in length, so must tolerate flow and load variations that directly reflect manufacturing operations. Most industrial operations use some chemicals that are toxic, corrosive or of extreme pH, hence it is inevitable that some of these chemicals arrive at the wastewater treatment plant. Small or continuous discharges usually represent only minor problems; however, accidental or deliberate release of such chemicals represents a major disruption for many wastewater treatment processes, particularly biologically based unit operations.

Thus, the industrial wastewater treatment plant must be able to accommodate a range of flow and load variations far greater than that encountered at municipal plants. Preferably data should be accumulated on a continuous or short time-interval basis, to indicate likely variability. A designer should also be aware of probable shock loads in the form of erratic discharges of specific chemicals or debris which may be discarded to the factory drainage system. Hence the pretreatment section of an industrial wastewater treatment plant is likely to include a screening operation and, often, facilities for flow balancing and neutralization.

Importance of wastewater treatment to industries

Centres of industrial activity were generated by the growth of manufacturing industry and the transfer of manpower from agriculture to manufacture in the Industrial Revolution. This process has accelerated and continues with the growth of the chemical and allied industries and the centralized processing of foodstuffs. Wastewaters from industry received relatively little treatment before the second half of the twentieth century. At this time, increased awareness of the limited supply of fresh water initiated a general campaign to reduce the pollutional load imposed upon this resource. As a consequence, regulatory agencies were set up to monitor and control discharges to fresh water and, subsequently, to other bodies of water. This led to a more stringent attitude towards industrial discharges to the sewers in order to control the loads at municipal wastewater treatment plants.

While responsible industrial organizations generally accept the basic premise that 'the polluter pays' and hence the manufacturer installs pollution control equipment, in many cases this represents an additional manufacturing cost. There are several examples of byproduct recovery to offset costs, or even to increase cash flow, but still most manufacturers see pollution control as a drain on resources. The capital cost of the wastewater treatment plant does not improve manufacturing efficiency and the plant incurs additional operating, maintenance and running costs. Therefore, while it is easy to advocate stringent water pollution control measures, these represent an economic disadvantage in many cases. The modification of some taxation principles to permit rapid depreciation of pollution control equipment — and even concessionary allowances on running costs for the equipment — can only be advantageous.

Until recently, industrial wastewater treatment plants were installed on the basis of lowest capital cost and were then expected to operate with minimum attention from unskilled or untrained operators. Such an approach, when combined with the erratic and difficult nature of many industrial wastewaters, has led to poor performance from many plants. For effective wastewater treatment the unit processes of pollution control must be integrated with the unit processes of manufacture and should be considered equally important to the overall efficiency of the industry. This ideal is rarely achievable, particularly in old-established industries where often the site has been reused and modified over several decades of industrial occupation. At such sites it may be difficult even to trace the drainage system or to separate waste streams.

In new industrial developments wastewater treatment should be considered at an early planning stage and fully integrated into the total facility. In some cases this can lead to major conflict between the wastewater treatment aspects and the manufacturing aspects of the industry. For example, in the processing of potatoes the use of a highly alkaline solution to remove skins — lye peeling — produces a strong solubilized waste that can be awkward to treat. More recent potato processing plants have replaced lye peeling, for some applications, with dry non-chemical peeling processes; while wastewater treatment is not the only consideration in making such process changes, it does have a significant influence on potato processing.

Another influence upon industry's attitude to pollution control is the potential volatility of any manufacturing activity. While municipal engineering tends to anticipate many decades, or even centuries, as the potential life of a given facility, manufacturing can change very rapidly as new products are developed or science and technology offer a novel approach to a product. Pollution control equipment must either be sufficiently flexible to accommodate such changes or must be expendable. Thus, while the majority of municipal wastewater treatment plants are built as massive structures of in-ground reinforced concrete, industrial plants tend to be of a less permanent nature and constructed, often above ground, of steel or glass-fibre.

Trends in industrial wastewater treatment

Pollution control is subject to short-term 'fashionable' changes, as are many other technologies. The installation of particular types of equipment — for example, surface or submerged aerators — or types of process — for example, biological or chemical — is influenced by a series of scientific, political and commercial factors. However, beyond these changes there are some perceptible trends, notably the recovery of byproducts, the integration of the wastewater treatment with manufacture, and the treatment and disposal of waste solids.

A major influence on many industrial activities has been the escalation of fuel costs and particularly of oil-based products. In several cases this has reduced the load to wastewater treatment plants. Many industrial processes produce 'waste oil' and when oil prices were low much of this was discharged into the drainage or sewer system. As the price of oil has increased, so has the incentive to recover or recycle oil, and the tendency to lose oil in wastewater has correspondingly decreased. Similarly, industries which rely heavily upon oil-based products — for example, the paint and solvent industries — have much greater incentives to recover losses of oil-based products; in this case the changes have been accelerated by more stringent air pollution legislation which has favoured water-based rather than solvent-based paints.

Higher energy costs have led to a re-examination of anaerobic processes for wastewater treatment. (Anaerobic treatment can represent a double saving: the process does not need to be supplied with oxygen, hence the power and operating costs of aeration can be eliminated; and methane can be generated as a potential supplementary fuel. The value of methane is limited unless the supply can be continuous and can be utilized at the existing site.) However, the mere conversion of carbon into a gas reduces the sludge mass for disposal, and this in itself can be a significant advantage. A disadvantage of anaerobic units has been that they were large, requiring many days retention and hence taking up an unacceptable proportion of potential manufacturing area. The development of high-rate contact anaerobic processes, such as the upflow blanket and the fluidized bed reactor, has overcome this major disadvantage and it can be assumed that anaerobic industrial wastewater treatment will continue to increase in importance.

The majority of wastewater treatment plants are for segregation rather than treatment of wastewaters. The raw wastewater is passed through a series of unit operations and is discharged as a less polluting, lower solids stream — the treated effluent. However, a high solids stream — waste sludge — remains and requires separate management. Some inorganic sludges, such as the mixed metal hydroxide sludges from plating works and the mercury-containing sludges from chlorine manufacture, contain toxic components and thus are unsuitable for direct disposal to landfill sites. For these intractable sludges chemical fixation methods have been devised to immobilize the species in a solid matrix. Such an option is relatively expensive but is a requirement for some waste solids. Organic sludges from biological processes are often a nuisance because of their large volume and tendency to become malodorous. In order to keep the size of a wastewater treatment plant to a minimum, aerobic processes with a high loading rate are installed and the resulting waste sludge is incompletely digested. The yield of sludge solids is often approximately 1 kg per kg BOD_5 removed and is wasted at a high moisture content (0.5–2% solids). The provision of sludge management for the wastewater treatment plant has to be considered as an integral part of the system rather than as a minor addition. The increasing costs of transporting waste sludges and more stringent regulation of disposal to landfill will increase the importance of sludge management.

Byproduct recovery, whether by converting high BOD effluents into methane gas or into a saleable product, or by recovering higher proportions of oils, solvents, metals or foodstuffs, must be advantageous. However, the potential for byproduct recovery is often limited by the mixed and erratic nature of the industrial wastewaters. Foodstuffs can be recovered from several food and beverage industries, but the discharge of chemicals such as detergents, alkalis, acids and disinfectants can devalue the product. Hence, while byproduct recovery is feasible in many new, well designed plants, it may be very restricted where the operation is long-established.

In fact, it is often found that major reductions in effluent loads can be achieved by simple flow inventories and improved 'housekeeping' rather than by very sophisticated recovery methods.

Conclusion

Industrial wastewaters are subject to wide variations of both flow and load. Moreover, rarely do any two factories produce similar effluents: no two tanneries, for instance, process the same mixture of hides by the same tanning methods, so their effluents cannot be similar. As manufacturing plants become older they tend to produce more polluted wastewaters as there is greater leakage from the manufacturing operations. The older manufacturing plants are likely to have been designed to use more water and power, to encourage the discharge of materials to the drainage system and to have been subject to modifications, all of which produce a wastewater which will be more costly to treat. These difficulties, combined with management attitudes which class pollution control equipment as 'non-productive', make it difficult to provide reliable effluent quality.

It is only by adopting an integrated approach to combine the manufacturing and pollution control operations that an industry can be fully optimized. In this way the requirements for effluent quality can be constructive, in that they encourage good housekeeping within the manufacturing area, provide an impetus to recycle and recover materials and minimize the environmental impact of the whole operation.

1 The management of wastewater from the petroleum refining industry

A S Vernick and P D Lanik, *Burns and Roe Industrial Services Corporation, Paramus, NJ*
B S Langer, *JRB Associates, Paramus, NJ*
S E Hrudey, *Department of Civil Engineering, University of Alberta*

1.1 Industry background

1.1.1 Production processes

In order to understand the petroleum refining industry from an environmental aspect, a rudimentary understanding of the industry from a process standpoint is necessary. A petroleum refinery is a complex combination of interdependent operations effecting the separation of crude molecular constituents, molecular cracking, molecular rebuilding and solvent finishing to produce an extremely wide variety of products (EPA, 1974a). Petroleum refining utilizes many different unit petrochemical processes to change crude oil physically and chemically in order to produce the various products in desired quantities. Refineries also incorporate ancillary facilities including storage tanks, electric power and steam generating facilities, as well as other support services.

Petroleum is a mixture of organic compounds and primary hydrocarbons that come from underground rock formations ranging in age from ten million to several hundred million years. The process by which it has formed and developed is not yet completely known (Bland and Davidson, 1967). The organic compounds that constitute crude oil can be physically separated through the application of heat in the distillation process. Separation results from the varying boiling ranges of the constituents of crude oil. Table 1.1 presents boiling-point data for a number of major petroleum products. Figure 1.1 presents a schematic of a simple petroleum distillation tower (Guthrie, 1960).

Although many major products can be produced from crude oil by simple physical separation processes, such as fractional distillation, the proportions of each product may not match the desired values, or the quality may not be adequate for the use intended. Therefore, many sophisticated chemical process operations also take place in a petroleum refinery, in order to produce the distribution, quality and quantity of products desired.

Figure 1.2 provides a simplified product schematic for a typical petroleum

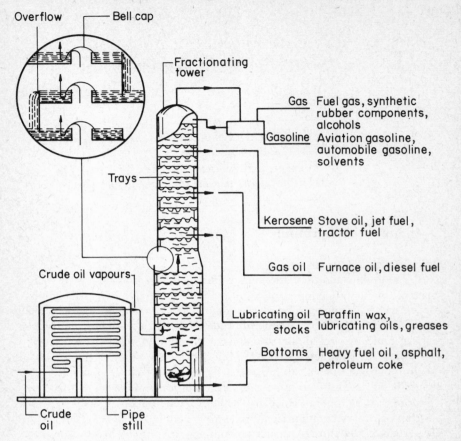

Fig. 1.1 Simple petroleum distillation tower

refinery (EPA, 1977). The diagram illustrates the interrelationships of the different categories of processes utilized and the products produced. The following paragraphs describe the basic elements of the key unit processes in a petroleum refinery.

Crude oil and product storage
Petroleum refineries require storage facilities for both crude oil and individual final products. The amount of storage required is quite variable depending on the type and reliability of crude supply and on the location and nature of markets. The crude oil storage area of a refinery serves to provide a working supply, equalize process flow and separate water and suspended solids from the crude oil.

During storage, water and suspended solids in crude oil and, in lesser quantities, in products tend to settle out to form a water layer at the tank

Table 1.1 Boiling ranges of petroleum products

Product	Boiling range (°C)
Liquefied petroleum gas	−44 to +1
Aviation gasoline	+32 to +149
Motor gasoline	+32 to +210
Jet propulsion fuel	+38 to +288
Cleaner's naphtha	+149 to +204
Kerosene	+177 to +288
Distillate fuel oil	+204 to +371
Refinery gas oil	+204 to +399+
Residual fuel oil	* to +399+

* Flash point usually specified.

bottom. This is typically in the form of a sludge which, in the case of crude oil, usually contains foul sulphur compounds and high dissolved solids concentrations.

Sludge withdrawn from storage tanks usually also includes some emulsified oil which is drawn off with the sludge. Tank cleaning operations, which are required intermittently, also constitute a significant source of oily wastewater from storage tanks. Wastewaters associated with crude oil and product storage are typically high in oil, suspended solids and COD.

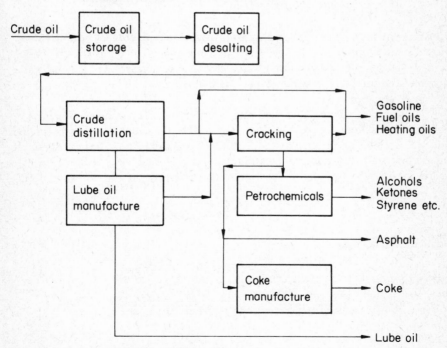

Fig. 1.2 Petroleum refining product manufacture

Crude oil desalting

The crude oil desalting process is a pretreatment step to remove the various dissolved salts contained in crude oil. This salt must be removed to avoid problems in subsequent processing operations. Problems which can be caused by salt include: scaling due to the precipitation of various ions; corrosion due to the presence of ions such as chlorides; and the poisoning of catalysts which are used in processes such as catalytic cracking and reforming.

Desalting is typically performed by either chemical or electrostatic desalters, although the latter method is becoming almost universally adopted. Chemical desalting (Fig. 1.3) utilizes a chemical demulsifying agent, a water–soda ash

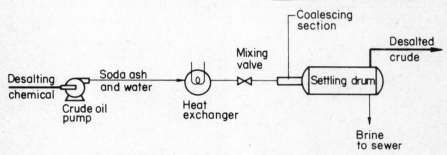

Fig. 1.3 Chemical desalting of crude oil

mixture, heat and thorough mixing to provide intimate contact between water and brine droplets (Vesik, 1974). The emulsion is then broken in a long coalescing section, which allows water droplets to coalesce. The oil and water layers separate in a settling drum, maintained at a higher temperature to promote separation.

As indicated, electrostatic desalting is becoming much more common than chemical desalting, because it does not require the addition of demulsifying chemicals. This process is illustrated in Fig. 1.4 (Vesik, 1974). As with chemical desalting, wash-water is applied and mixed with crude oil at elevated temperatures, to provide thorough contact between entrained salt and the wash-water.

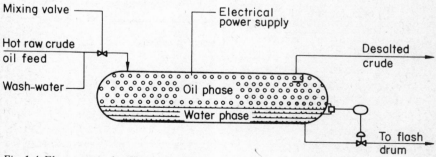

Fig. 1.4 Electrostatic desalting of crude oil

In this case, a water-in-oil emulsion is formed. This emulsion is destabilized by the application of an electrostatic field to the mixture, which causes the water droplets to agglomerate and separate.

In both instances, the wastewater produced by desalting contains emulsified and free oils, ammonia, phenols, sulphides, suspended solids and dissolved solids.

Crude distillation

Distillation is the basic refining process for the separation of crude petroleum into intermediate fractions of specified boiling-point ranges. This separation or fractionation takes advantage of the differing boiling points and vapour pressures of the various components in the crude oil mixture.

Atmospheric distillation is the fundamental process common to all petroleum refineries, although the arrangement of sub-processes before and after the atmospheric distillation unit may vary. The heart of this process is the fractionating column. This column is packed with a series of trays which allow vapour to pass up the column but also collect condensed liquid and allow it to run back down the column, or reflux (Fig. 1.5; Vesik, 1974). The provision of reflux is very important to the functioning of the column. If vapours were drawn off without reflux, the degree of separation achieved at a given yield would be much lower.

Steam is injected at the bottom of the column and rises up through the tower. This steam maintains a high temperature in the bottom section of the column and, because steam is immiscible with the hydrocarbons, it has the effect of increasing the total vapour pressure of the mixture. This results in the stripping of the lighter ends from the condensate which is falling down the tower. Thus, the residue drawn off the bottom of the column should be free of any lighter ends. These stripped components are then carried back up the column to find their correct condensation stage. Steam stripping is also applied to the various vapour fractions being drawn off the column, to return the lighter fractions to the tower while collecting and condensing the desired fraction.

The steam applied at the various stages to the process is in direct contact with hydrocarbons. It is eventually carried over with various fractions and is separated out by gravity when the fraction is condensed. These steam condensates are invariably foul, and constitute a foul or sour condensate waste stream, containing sulphides, ammonia, chlorides, mercaptans and phenols.

In addition to the atmospheric distillation process, it is normally necessary to subject the residual or bottoms from atmospheric distillation to a second and/or third stage distillation, conducted under vacuum. The vacuum is usually produced by steam jet ejectors, which are essentially venturi jets with a connection to the throat of the jet that draws vapour from the connected medium. Obviously, the vapour stream drawn into the jet will contaminate the steam and produce a contaminated condensate. The steam applied for steam stripping in vacuum columns also produces a contaminated condensate.

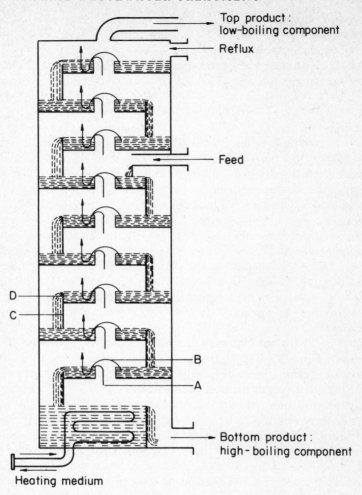

Fig. 1.5 Bubble tray fractionating column
A   Vapour   C   Downcomer
B   Bubble-cap   D   Weir

The vapours may be withdrawn by the injector into a barometric condenser. In this device, cooling-water is allowed direct contact with the vapour to be condensed, which results in a large volume of contaminated wastewater. These wastewaters exhibit a very stable oil-in-water emulsion, which is subsequently difficult to separate. Alternatively, surface condensers (shell-and-tube heat exchangers) may be used, which isolate cooling-water from the condensing vapours. When these are employed, the wastewater produced by vacuum distillation is limited to condensate from steam stripping and the steam jet ejectors.

Cracking

In this process, heavy oil fractions are converted into lower-molecular-weight fractions including domestic heating oils, high octane gasoline stocks and furnace oils. The cracking process increases the yield of gasoline taken from the crude oil and improves its quality. By using cracking, refiners can double their gasoline output per m^3 of crude oil charged to their distillation towers or stills. Three types of cracking are used: thermal, catalytic and hydrocracking.

Thermal cracking is a process for breaking down the larger molecules in heavy gas oils into lower-molecular-weight fractions strictly by the application of heat (Fig. 1.6; Orr, 1974). Thermal cracking is accomplished by heating

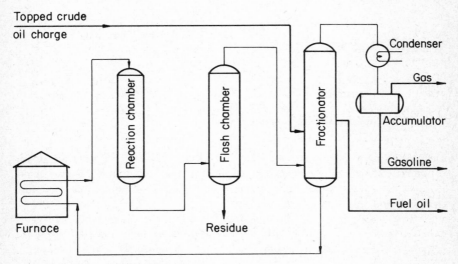

Fig. 1.6 Thermal cracking unit

(480°–603°C) at 42–69 atm without the use of a catalyst. The main waste streams from the process are due to condensate from steam stripping in the fractionating column. This steam condensate collects in the overhead accumulator and separates by gravity from the hydrocarbon products. This condensate typically contains ammonia, phenols and sulphides, is high in BOD, COD and alkalinity and is therefore classified as a sour condensate.

Catalytic cracking accomplishes the cracking step at milder conditions than necessary for thermal cracking, by the use of a catalyst. It is an important process at any integrated refinery for the production of large volumes of high octane gasoline stocks, furnace oils, and other middle-molecular-weight distillates. The most common version of this process is fluid bed catalytic cracking (Fig. 1.7; Orr, 1974). This process uses a finely powdered catalyst in a fluidized bed reactor. The catalysts commonly used are silica- or alumina-treated bentonite, fuller's earth, aluminium hydrosilicates, bauxite and zeolite.

The 'cat cracker' is usually the largest single source of sour and phenolic

8 INDUSTRIAL WASTEWATER TREATMENT

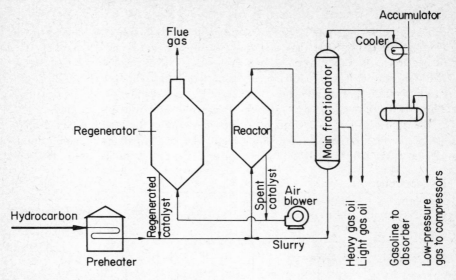

Fig. 1.7 Fluid bed catalytic cracking unit

wastewater in a large refinery. The wastewater is derived from the steam condensate from the overhead accumulator and condensate from steam stripping of side-streams. The major pollutants are oil, sulphides, phenols, ammonia and traces of cyanides.

Hydrocracking is a catalytic cracking process in the presence of hydrogen and has greater flexibility in adjusting operations to meet changing product demands. By providing hydrogen, hydrocracking operates at lower temperatures, but higher pressures, than normal fluid bed catalytic cracking. The product is similar in molecular weight distribution to that achieved by catalytic cracking, but the formation of olefins is reduced. Like catalytic cracking, this process results in sour condensates from steam stripping in the fractionator. Because of the tendency for hydrogen to strip sulphur from hydrocarbon mixtures, the condensates may be expected to be higher in sulphide content.

Hydrocarbon rebuilding

Higher octane products for use in gasoline may be manufactured by two hydrocarbon rebuilding techniques: polymerization or alkylation. In these processes, small hydrocarbon molecules are combined into large molecules: the reverse of cracking. The resulting products are valuable components of high-quality motor fuel and aviation gasolines.

The purpose of polymerization is the conversion of light olefin feedstocks to higher octane polymers. The polymerization step itself does not produce any effluent, but sulphur removal, typically by caustic washing, is required before charging to this process. This operation produces sour water, high in sulphides, mercaptans, ammonia, suspended solids and oils. Polymerization units are generally small in existing refineries, and the process is not commonly being incorporated into new refineries.

Alkylation is a reaction of an isoparaffin such as isobutane and an olefin (propylene, butylene) in the presence of an acid catalyst to produce larger branched-chain compounds (alkylate). The catalyst used is either sulphuric acid or hydrofluoric acid. The main waste stream from sulphuric acid alkylation is the spent caustic and the caustic-contaminated wash-water. This spent caustic is not usually high in sulphides, most of the sulphur present being derived from the sulphuric acid. The hydrofluoric acid process does not produce spent acid or spent caustic streams. These plants are equipped with hydrofluoric acid recovery units. The major source of wastewater is the overhead accumulator on the fractionator. Both hydrofluoric acid and sulphuric acid alkylation units require containment systems to prevent acid spills from reaching the sewer.

Hydrocarbon rearrangements

Isomerization and reforming are two process techniques for obtaining higher octane gasoline blending stock. Isomerization, a molecular rearrangement process rather than a decomposition process, generates no major pollutant discharge. Reforming, a mild decomposition process, is used to convert low octane naphtha, heavy gasoline and naphthene-rich stock to high octane gasoline (aromatics) plus isobutane, with hydrogen as a byproduct.

The catalyst used in reforming is usually in the form of small-diameter spherical balls of alumina coated with platinum. This catalyst is very susceptible to poisoning by sulphur compounds, so the feedstock must be hydrotreated to remove sulphur. Reforming produces a reduction in the average molecular size and boiling range of the hydrocarbon feed. This process is relatively clean, generating low volume, slightly alkaline discharges with small quantities of sulphides, ammonia, mercaptans and oil present.

Hydrotreating

Hydrotreating processes are used to purify and pretreat various feedstocks by reacting with hydrogen. Product contaminants, including sulphur and nitrogen compounds, odour, colour and gum-forming materials, are removed in this process.

Many different hydrotreating processes are used, depending on the feedstock and intended use of the product. Common applications include:

- Pretreatment of reformer feedstock
- Naphtha desulphurization
- Lube oil polishing
- Pretreatment of cat-cracking feedstock
- Heavy gas oil and residual desulphurization
- Naphtha hydrogenation

The strength and quantity of wastewaters generated by hydrotreating is largely dependent upon the specific process and feedstock used. Wastewaters are derived as condensates from fractionating the product hydrocarbons and are mainly contaminated by ammonia and sulphur compounds. Phenol may also be present in the wastewater.

Solvent refining

Various chemicals and solvents are used to improve the quality of a particular feedstock component. The compounds removed or isolated by this process may be highly objectionable in the specific product being prepared, but may be desirable in making other products or may be converted into desirable materials. The major pollutants from solvent refining are the solvents themselves, many of which can produce a high BOD. Under ideal conditions the solvents are continually recirculated, but in practice some solvent is always lost, usually through leaks at pump seals and flanges. Oil and solvent are major wastewater constituents.

Asphalt production

The reduced crude fraction or residual taken from the bottom of the vacuum still may be blended into heavy fuel oil or may be made into asphalt by oxidation in an asphalt still. An asphalt still attempts to improve the physical properties of asphalt by oxidizing soft tar to strip off hydrogen and water. The hydrocarbons stripped of oxygen will react with one another to form heavier molecules which are more suitable for asphalt.

Asphalt feedstock is contacted with hot air at $203°-280°C$ in a tall reactor column to obtain a desirable asphalt product. Gas, light oil vapour and steam pass out at the top of the column to a condensing tank. Wastewater is derived from steam added to the reactor for stripping volatiles, as well as a small quantity of water produced from oxidation reactions with the asphalt. The water separated out is very oily, high in BOD and usually sour as a result of the normally high sulphur content of the residual.

Lubricating oil manufacture

Lubricating oils require closely controlled properties and are generally only manufactured from special high-grade feedstocks. However, even with high-grade feedstocks, lube oils must be treated to remove asphalt, wax and hydrocarbons whose viscosity is temperature sensitive (generally aromatic compounds).

Lubricating oil manufacturing encompasses a number of unit processes including hydrofining, hydrofinishing, propane dewaxing and fractioning, white oil manufacture, solvent treating and extraction, duotreating, oil fractionation, bright stock treating, centrifuging, chilling, clay contacting and percolation, acid treatment, phenol extraction, lube and fuel additive operations, petroleum oxidation, manufacturing of grease and allied products, blending, product finishing, etc.

Once the lube oil has been deasphalted and dewaxed, it often requires finishing treatment for the removal of colour and other impurities. The two processes used are clay and acid treatment. In clay treatment, the lube feed is mixed with clay to form a slurry, heated, and then vacuum filtered to produce clarified oil. In acid treatment the lube feed is acid extracted, then water washed. This operation produces acidic rinse-waters and acid sludges for disposal, which are high in dissolved and suspended solids, sulphates and sulphonates and which form stable oil emulsions. Clay treatment, on the other hand, produces only small volumes of wastewater, with clay and free and emulsified oils the major pollutants. Acid and clay treatments are being replaced by hydrotreating in many cases and, in particular, acid treating of lube stocks is being phased out of operation.

Product finishing
Product finishing is a broad process category covering drying and sweetening, which includes the removal of impurities such as sulphur, water and trace impurities from gasoline, kerosene, jet fuels, domestic heating oils and other middle distillates.

Sweetening refers to the removal of hydrogen sulphide, mercaptans and various thiophenes. This is usually done by oxidation of mercaptans and disulphides, removal of mercaptans or destruction and removal of all sulphur compounds. Drying is done by salt filters, absorptive clay beds or electrostatic water separation.

The most common chemical process for sweetening is caustic washing. This involves washing the product with dilute sodium hydroxide to remove acid sulphur compounds, such as hydrogen sulphide and mercaptans. After the product has been mixed with the caustic wash solution it is allowed to coalesce, aided by electrostatic precipitation, and spent caustic solution is discharged. The spent caustic may be predominantly phenolic or sulphidic, depending upon the contaminants in the product.

Blending and packaging
Blending and packaging operations are normally the final processing step prior to shipment to markets. Normally the largest volume operation is the blending of gasoline stock. Gasoline is usually a blend of butane, straight run naphtha, cat-cracked gas oils, alkylate and reformate, plus additives such as tetraethyl lead, deicing agents, detergents, etc. Blending is accomplished in a blending header, fed by individual component feeds under close flow control.

Additives are included in small proportions in a wide variety of products to improve certain of their characteristics. Examples are the use of additives to improve antiknock characteristics of motor and aviation gasolines, or to improve the oxidation stability of lubricating oils.

Blending of various feedstocks with additives and packaging of the products are relatively clean processes, provided that spills are prevented or adequately contained. The primary source of wastewater is the washing of railroad tank cars or tank trucks, which produces wastewaters with high concentrations of emulsified oil. Spills commonly occur at asphalt plants when hot asphalt is loaded into tanks containing residual water: the asphalt boils over, and spills result.

Production of petrochemicals
These operations are extremely varied, and include production of a wide range of products such as alcohols, ketones, cumene, styrene, benzene, toluene, xylene, olefins, cyclohexane, etc. Many petrochemicals are manufactured directly, while others are derivatives from intermediate products. Petrochemicals manufacture is an increasingly important aspect of the petroleum processing industry. Wastewaters from these processes are quite variable and dependent upon the specific operation employed. These problems are addressed in Chapter 3.

Utilities
Most refineries generate their own steam for use in the processes described above, although some plants buy steam and electricity from outside sources. Those refineries which generate their own steam usually require extensive water treatment facilities in order to satisfy the high purity requirements of the feed-water to the boilers. Chemical softening and ion-exchange demineralization are commonly used, giving rise to waste sludges and regeneration brines for disposal.

Non-contact steam is usually recirculated to the boiler house, although inevitable losses occur as a result of leakage and contamination by leaks in heat exchangers. Because the dissolved solids tend to build up with recycle, a portion of the recycle must be discharged as boiler blowdown. Boiler blowdown is relatively free from process contaminants but contains dissolved solids and treatment additives. This stream typically amounts to 5–10% of the total steam generated.

The steam applied to contact-process uses forms a major source of process wastewater when condensed. As noted previously, a large number of these condensates are sour in nature.

Cooling-waters are required throughout the process and power generation facilities. Both once-through and recirculated cooling-water systems are in use. Generally, once-through systems are no longer adopted in petroleum refineries because of the excessive water requirement and the possibility of significant mass discharge of pollutants at low concentrations (but high volume) due to leaking heat exchangers.

Recirculated systems require the discharge of heat prior to recirculation. This is commonly done in cooling towers where the hot water is cascaded down a baffled tower provided with an updraft of air. A small amount of water evaporates ($\simeq 1\%$ evaporation loss per 5.5°C of cooling) and some is lost due to drift, the carryover of small water drops in the updraft. Finally, in order to maintain solids contents at < 3–4 times the intake water concentration, there is a requirement for a blowdown of 0.3% per 5.5°C of cooling.

Like boiler feed-waters, cooling-waters must be treated to avoid scaling and corrosion and also to control slime growth. Typically, scaling may be controlled by adjusting the alkalinity with sulphuric acid. Corrosion is controlled by the use of zinc or chromate salts, phosphates or polyphosphates. Slime growth may be controlled by chlorinated phenols, quaternary ammonium compounds, copper salts or shock chlorine treatment. In most cases, shock chlorine treatment is the method of choice.

Process summary

The unit processes described are used in varying combinations in petroleum refineries throughout the industry. Generally, a simple petroleum refinery includes catalytic reforming and treating processes in addition to crude oil distillation. A more complex refinery also includes catalytic cracking, polymerization, alkylation and asphalt oxidation as well as other selected unit processes. A very complex refinery may include high vacuum fractionation, solvent extraction, deasphalting, dewaxing and treating processes, in addition to those found in simple and complex refineries (EPA, 1974a).

1.1.2 The products

The products of the petroleum refining industry range from gasoline for automobiles to organic chemical products for the pharmaceutical industry. Specific individual products are far too numerous to delineate here, but the following list summarizes the major product areas (Nelson, 1969):

- Volatile products — liquefied gases and natural gasoline
- Light oils — gasolines, rocket and jet fuels, solvents, tractor fuel, kerosene
- Distillates — range oil, furnace distillates, diesel fuel, gas oil
- Lubricating oils — motor, engine, machine, cylinder, spindle, gear, etc., oils
- Greases and waxes — paraffin wax, microcrystalline wax, petrolatum, salve bases, greases
- Residues — fuel oil, coke, asphalt, carbon black, etc.
- Specialties — medicinal products, hydrocarbons, chemicals, insecticides, etc.

Thus many items that we use or come into contact with every day have their origins in a petroleum refinery. Many of the products, which are normally referred to as second-generation petrochemicals, are used both as final products and as raw materials for the organic chemical industry. Table 1.2 lists the major petrochemicals produced from crude petroleum and natural gas.

Table 1.2 Petrochemical products from petroleum and natural gas

AROMATICS

Distillates and solvents
Benzene
Cresylic acid, crude
Naphthenic acids
Toluene
Xylenes

ALIPHATICS

Methane
Ethane
Ethylene
Propane
Propylene and C_3 mixtures
1,3-Butadiene
n-Butane
Butylenes
Isobutane
Isobutylene
C_5 hydrocarbons
Diisobutylene
Dodecene
Nonene
Derivatives

These products are generally manufactured in only the very sophisticated and large oil refineries.

1.1.3 Profile of the industry

The refining of petroleum is conducted worldwide, the concentration of refining capacity generally being found in the industrialized nations where demand for energy and consumer products is the greatest. In the mid-1970s, the refining capacity of the non-Communist world approximated 0.95 million $m^3 d^{-1}$, with the USA, Canada, Japan and Western Europe accounting for more than two-thirds of that total. This volume of crude oil was processed through 730 plants, of which approximately 300 were in the USA and Canada, 160 were in Western Europe and 50 were located in Japan. The remaining 220 refineries were scattered throughout Latin America, Africa, the Middle East and the Far East (Anon., 1977).

The trend in the industry throughout the world has generally been one of restricted growth, due primarily to price levels and energy conservation. However, in the USA expansion of refining capacity continued unabated throughout the 1970s. Since January 1974, USA crude capacity has grown some 25% compared with only 4% in the European Economic Community (EEC) and 11% in Japan.

The US Department of Energy reports that over 64 new refineries, with almost 0.16 million m^3 d^{-1} of capacity, have been built in the USA since 1974. This construction activity resulted in expansion in the number of active refineries in the USA to about 300 in 1980, with a total capacity of almost 0.286 million m^3 d^{-1}. The average size of USA refineries is about 9500 m^3 d^{-1}. In contrast, the EEC has five fewer refineries than in 1974, but the average size is approximately 4750 m^3 d^{-1} (Anon., 1980).

Within the USA, refineries are concentrated in areas of major crude production (California, Gulf Coast Region) and major population areas (East Coast and Midwest). Forty-one states have at least one plant, refineries also being located in Puerto Rico, the Virgin Islands and Guam.

Age of a refinery is very difficult to determine. Refineries are constantly upgrading their equipment and expanding their facilities. A plant that was initially constructed in 1925 may consist of mostly new equipment today, with only some crude distillation units surviving from the original plant. These new units and brand new refineries have generally been designed and constructed with water conservation and pollution control as prime objectives. Consequently, wastewater discharges from these facilities typically are smaller in volume, and also contain a lower pollutant loading per m^3 of crude throughput, than effluents originating from older facilities.

1.2 Wastewater characteristics

Refinery wastewater is a general term used to describe the various wastewaters discharged from a refinery complex. These wastewaters include, depending upon the individual plant characteristics, stormwater, ballast-water, sanitary wastewater, once-through non-contact cooling-water and process wastewater.

This section addresses only process wastewater, as it is the most significant discharge on a consistent day-to-day basis. However, this does not imply that the other sources of refinery wastewater mentioned above are not of concern. Refinery stormwater or, more precisely, the plant runoff from precipitation can be contaminated by raw materials or products. Stormwater pollutants typically include oil and grease. Control measures for stormwater pollution include simple good-housekeeping measures, storm sewer segregation, and storm retention facilities (holding ponds).

Ballast-water, used at coastal refineries, is the water taken on by seagoing vessels to improve their stability. Typically, this water is contaminated by the previous contents of the compartment in which it is held. Where ballast-water dumping at sea is not feasible or not permitted, this water is discharged to the refinery for treatment, normally by gravity separation.

Sanitary wastewater in a refinery typically has the pollutant characteristics of domestic sewage. Treatment is normally by biological oxidation. The refinery may discharge this wastewater directly to a municipal sewerage system or include it with process wastewater for treatment on site.

Once-through non-contact cooling-water *should not* contain pollutants. However, the potential for leakage in the system does exist and, when detected, can be controlled by normal repairs.

1.2.1 Sources of process wastewater

Process wastewater is discussed here in terms of its sources, raw quality and quantity, and treated effluent quality. The information presented here is based on USA refinery industry data for the years 1976–1978 (EPA, 1979).

The sources of process wastewater within a refinery include the following:

- Non-segregated cooling water
- Cooling tower blowdown
- Boiler blowdown
- Oily process water
- Sour water
- Water treatment system blowdown
- Air pollution control equipment blowdown

Non-segregated cooling-water is once-through cooling-water which has been combined with other wastewater within the refinery. This is a typical source of process wastewater in old refineries, where the cooling-water has become contaminated in the cooling process. Contamination typically includes a low concentration of oil and grease. However, because of the large volume of flow, the mass of pollutants can be substantial. In general, once-through cooling is being replaced in the industry by more efficient methods.

Cooling tower blowdown is the concentrated wastewater discharge from the cooling tower water cycle. Blowdown from the system is necessary to eliminate solids which are built up by the evaporative process. The characteristics of blowdown include relatively low volume and high concentration of dissolved solids, plus concentration of the water treatment chemicals which are added to inhibit corrosion and the formation of algae in the system. Water treatment additives include materials such as chrome, zinc, chlorine and biocides. Cooling towers are common in the industry, particularly in areas where an abundant source of fresh water for once-through cooling is not available.

Boiler blowdown is the concentrated wastewater discharge from a boiler-type heating and steam-generating system. The discussion presented above on cooling tower blowdown, regarding the need for blowdown and its characteristics, applies to boiler blowdown as well. However, the volume of boiler blowdown is typically small compared with that of cooling towers. Boiler blowdown, as a source of process wastewater, is common in the refining industry.

Oily process wastewater is defined here to include the following:

- Barometric condenser blowdown: the wastewater produced as a result of vapour cooling
- Contact process water: water contaminated by process utilization
- Non-sour process condensate: the condensate produced in process operations which is not highly contaminated by ammonia and hydrogen sulphide
- Vent scrubber water: the wastewater generated in the control, or scrubbing, of process vent gases
- Tank drainage: the waste material from storage facilities which is collected in the refinery's sewer system
- Flare seal water blowdown: the wastewater from the liquid seal for the waste gas combustion
- Laboratory drainage: the waste produced in the analytical operations of product testing and control
- Maintenance decontamination water: the wastewater generated in the cleanup operations associated with equipment maintenance
- Pad wash-water: the washdown wastewater generated in the area of equipment pads.

A refinery typically generates oily process wastewater that is some combination of the individual sources listed above.

Sour water is process wastewater which is contaminated with ammonia and hydrogen sulphide. This is a common type of wastewater in a refinery and generally represents between 8 and 18% of the total process wastewater generated. Sources of sour water include the crude desalter, the crude unit and the desulphurization, catalytic cracking and gas recovery processes.

Water treatment system blowdown is the wastewater discharged from a refinery's fresh water treatment facility. The need for water treatment is dependent on the quality of fresh water available to the refinery.

Air pollution control equipment blowdown is the wastewater discharge from exhaust gas scrubbers in refineries which require control of emissions from combustion processes. This is a source of wastewater in larger refineries.

1.2.2 Quality of process wastewater

Typically, a refinery recovers oil from the process wastewater. A gravity-type separation unit such as an American Petroleum Institute (API) separator is normally used in this recovery process. The effluent water from the oil recovery unit is considered to be the plant's raw process wastewater. The quality of this wastewater is addressed below in terms of traditional pollutant parameters, as well as toxic or priority pollutants.

Traditional pollutant parameters

Raw process wastewater quality can vary widely from plant to plant, depending on process characteristics and operating methods. Table 1.3 defines subcategories used in Tables 1.4 and 1.5 (EPA, 1980), which describe the

Table 1.3 Process subcategories (EPA, 1980)

Subcategory	Basic refinery operations included
Topping	Topping and catalytic reforming, whether or not the facility includes any other process in addition to topping and catalytic process. This subcategory is not applicable to facilities which include thermal processes (coking, visbreaking, etc.) or catalytic cracking
Cracking	Topping and cracking, whether or not the facility includes any processes in addition to topping and cracking, unless specified in one of the subcategories listed below
Petrochemical	Topping, cracking, and petrochemical operations, whether or not the facility includes any process in addition to topping, cracking, and petrochemical operations,[a] except lube oil manufacturing operations
Lube	Topping, cracking and lube oil manufacturing processes, whether or not the facility includes any process in addition to topping, cracking and lube oil manufacturing processes, except petrochemical operations[a]
Integrated	Topping, cracking, lube oil manufacturing processes and petrochemical operations, whether or not the facility includes any processes in addition to topping, cracking, lube oil manufacturing processes and petrochemical operations[a]

[a] The term 'petrochemical operations' means the production of second-generation petrochemicals (i.e., alcohols, ketones, cumene, styrene, etc.) or first-generation petrochemical and isomerization products (i.e., BTX, olefins, cyclohexane, etc.) when 15% or more of refinery production is as first-generation petrochemicals and isomerization products.

characteristics of this wastewater on the basis of concentration and mass, respectively.

Toxic pollutant parameters
Raw process wastewater quality is addressed here in terms of toxic pollutants, designated 'priority pollutants' by the US Environmental Protection Agency. These pollutants are grouped as volatile organics, semi-volatile organics, pesticides and metals; Table 1.6 gives the US EPA list and also typical concentrations of toxic pollutants likely to be detected in refinery wastewaters.

1.2.3 Quantity of process wastewater

Figure 1.8 illustrates the typical generation of process wastewater in USA refineries. In a particular refinery, the quantity of process wastewater generated can vary considerably from these typical values depending on process characteristics and operating methods. In refineries with a crude throughput of up to $1600 \, m^3 \, d^{-1}$, the variation can be from 0.001 to 4.5 times the typical values given in the figure. Similarly, refineries in the 1600 to $8000 \, m^3 \, d^{-1}$ range can have a variation factor of 0.01 to 6. Refineries in the size range of 8000 to $32\,000 \, m^3 \, d^{-1}$ have, in comparison with smaller refineries, a more narrow range of variation from the typical wastewater generation values given in Fig. 1.8. These plants can have a variation factor of 0.06 to 2.8. Finally, refineries

Table 1.4 Raw wastewater characterization by subcategory in petroleum refining (concentrations in mg l^{-1}) (EPA, 1980)

Characteristics	Topping subcategory		Cracking subcategory		Petrochemical subcategory		Lube subcategory		Integrated subcategory	
	Range	Median	Range	Median	Range	Median	Range	Median	Range	Median
BOD$_5$	10–50	23.3	30–600	138	50–800	144	100–700		100–800	114
COD	50–150	107	150–400	383	300–600	418	400–700		300–600	261
TOC	10–50	20	50–500	66.3	100–250	135	100–400		50–500	51.1
TSS	10–40		10–100		50–200		80–300		20–200	
Nitrogen, as ammonia	0.05–20	2.72	0.5–200	28.6	4–300	42.1	1–120		1–250	14.5
Phenolic compounds	0–200	0.80	0–100	6.04	0.5–50	10.0	0.1–25		0.5–50	2.25
Sulphides	0–5	0.240	0–400	1.24	0–200	176	0–40		0–60	1.24
Oil and grease	10–50	25	15–700	52.8	20–250	44.9	40–400		20–500	44.1
Total chromium	0–3	0	0–6	0.109	0–5	0.471	0–2		0–2	0.272

Table 1.5 Raw wastewater[a] loadings by subcategory in petroleum refining (net kg per 1000 m³ of feedstock) (EPA, 1980)

Characteristics	Topping subcategory		Cracking subcategory		Petrochemical subcategory		Lube subcategory		Integrated subcategory	
	Range[b]	Median	Range[b]	Median	Range[b]	Median	Range[b]	Median	Range[b]	Median
Flow[c]	8.00–558	66.6	3.29–2750	93.0	26.6–443	109	68.6–772	117	40.0–1370	235
BOD$_5$	1.29–217	3.43	14.3–466	72.9	40.9–715	172	62.9–758	217	63.5–615	197
COD	3.43–486	37.2	27.7–2520	217	200–1090	463	166–2290	543	72.9–1490	329
TOC	1.09–65.8	8.01	5.43–320	41.5	48.6–458	149	31.5–306	109	28.6–678	139
TSS	0.74–286	11.7	0.94–360	18.2	6.29–372	48.6	17.2–312	71.5	15.2–226	59.1
Sulphides	0.002–1.52	0.054	0.01–39.5[d]	0.94[d]	0.009–91.5	0.86	6.5–96.2	24.1		
Oil and grease	1.03–88.7	8.29	2.86–365	31.2	12.0–235	52.9	4.58–52.9	8.29	0.61–22.6	3.78
Phenols	0.001–1.06	0.034	0.19–80.1	4.00	2.55–23.7	7.72	1×10^{-5}–20.0	0.014	0.52–7.87[d]	2.00[d]
Ammonia	0.077–19.5	1.20	2.35–174	28.3	5.43–206	34.3	23.7–601	120	20.9–269	74.9
Chromium	0.0002–0.29	0.007	0.0008–4.15	0.25	0.014–3.86	0.234	0.002–1.23	0.046	0.12–1.92	0.49

[a] After refinery API separator.
[b] Probability of occurrence ⩽ 10% or 90% respectively.
[c] 1000 m³ per 1000 m³ of feedstock throughput.
[d] Sulphur.

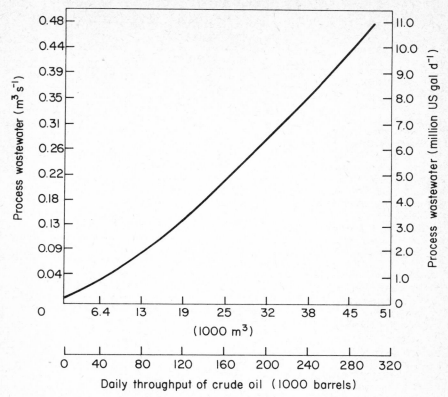

Fig. 1.8 Typical process wastewater generation rates

greater than 32 000 m³ d⁻¹ have the least range of variation — within 0.2 to 2.3 times the typical values given in Fig. 1.8.

1.2.4 Quality of treated effluent

API separator effluent, or raw process wastewater, discussed above, must usually be treated prior to discharge into a receiving water. Discharge standards and technology that apply to this treatment are discussed in the following sections of this chapter. Treated effluent, as discussed here, is process wastewater that has been biologically treated, by means of activated sludge, aerated lagoons or rotating biological contactors, and then polished in final sedimentation ponds or by filtration.

In addition to the factors mentioned above that affect the quality of raw process wastewater, treated effluent quality also depends on the operation and performance of the end-of-pipe treatment system. With this in mind, the quality of treated effluent is addressed below in terms of traditional pollutant parameters, as well as toxic or priority pollutants.

Table 1.6 Toxic pollutants

US EPA priority pollutants	Pollutants typically detected in raw process wastewater	Typical concentration[a] ($\mu g\, l^{-1}$)
VOLATILE ORGANICS		
Acrolein		
Acrylonitrile		
Benzene	Benzene	430
Carbon tetrachloride		
Chlorobenzene		
1,2-Dichloroethane		
1,1,1-Trichloroethane		
1,1-Dichloroethane		
1,1,2-Trichloroethane		
1,1,2,2-Tetrachloroethane		
Chloroethane		
Bis-(chloromethyl) ether		
2-Chloroethyl vinyl ether		
Chloroform		
1,1-Dichloroethylene		
1,2-trans-Dichloroethylene		
1,2-Dichloropropane		
1,3-Dichloropropylene		
Ethylbenzene	Ethylbenzene	130
Methylene chloride		
Methyl chloride		
Methyl bromide		
Bromoform		
Dichlorobromomethane		
Trichlorofluoromethane		
Dichlorodifluoromethane		
Chlorodibromomethane		
Tetrachloroethylene		
Toluene	Toluene	1300
Trichloroethylene		
Vinyl chloride		
SEMI-VOLATILE ORGANICS		
Acid-extractable fraction		
2,4,6-Trichlorophenol		
Parachlorometa cresol		
2-Chlorophenol		
2,4-Dichlorophenol		
2,4-Dimethylphenol	2,4-Dimethylphenol	80
2-Nitrophenol		
4-Nitrophenol		
2,4-Dinitrophenol		
4,6-Dinitro-o-cresol		
Pentachlorophenol		
Phenol	Phenol	1500
Base-neutral fraction		
Acenaphthene	Acenaphthene	270
Benzidine		
1,2,4-Trichlorobenzene		
Hexachlorobenzene		

Table 1.6 Toxic pollutants (cont.)

US EPA priority pollutants	Pollutants typically detected in raw process wastewater	Typical concentration[a] ($\mu g\ l^{-1}$)
Hexachloroethane		
Bis-(2-chloroethyl) ether		
2-Chloronaphthalene		
1,2-Dichlorobenzene		
1,3-Dichlorobenzene		
1,4-Dichlorobenzene		
3,3'-Dichlorobenzidine		
2,4-Dinitrotoluene		
2,6-Dinitrotoluene		
1,2-Diphenylhydrazine		
Fluoranthene	Fluoranthene	7
4-Chlorophenyl phenyl ether		
4-Bromophenyl phenyl ether		
Bis-(2-chloroisopropyl) ether		
Bis-(2-chloroethyoxy) methane		
Hexachlorobutadiene		
Hexachlorocyclopentadiene		
Isophorone		
Naphthalene		
Nitrobenzene		
n-Nitrosodimethylamine		
n-Nitrosodiphenylamine		
n-Nitrosodi-N-propylamine		
Bis-(2-ethylhexyl) phthalate		
Butyl benzyl phthalate		
Di-n-butyl phthalate		
Di-n-octyl phthalate		
Diethyl phthalate		
Dimethyl phthalate		
1,2-Benzanthracene		
Benzo (A)pyrene		
3,4-Benzofluoranthene		
11,12-Benzofluoranthene		
Chrysene	Chrysene	15
Acenaphthylene		
Anthracene		
1,12-Benzoperylene		
Fluorene		
Phenanthrene	Phenanthrene	170
1,2:5,6-Dibenzanthracene		
Indeno(1,2,3-C,D) pyrene		
Pyrene		
PESTICIDES		
Aldrin	Pesticides are not normally associated with refinery process operations	
Dieldrin		
Chlordane		
4,4'-DDT		
4,4'-DDE		
4,4'-DDD		
Alpha-endosulphan		

Table 1.6 Toxic pollutants (cont.)

US EPA priority pollutants	Pollutants typically detected in raw process wastewater	Typical concentration[a] ($\mu g\, l^{-1}$)
Beta-endosulphan		
Endosulphan sulphate		
Endrin		
Endrin aldehyde		
Heptachlor		
Heptachlor epoxide		
Alpha-BHC		
Beta-BHC		
Gamma-BHC		
Delta-BHC		
PCB-1242		
PCB-1254		
PCB-1221		
PCB-1232		
PCB-1248		
PCB-1260		
PCB-1016		
Toxaphene		
TCDD		
METALS		
Antimony		
Arsenic	Arsenic	120
Beryllium		
Cadmium		
Chromium	Chromium	700
Copper	Copper	65[b]
Cyanide	Cyanide	150[b]
Lead	Lead	140[b]
Mercury		
Nickel	Nickel	20[b]
Selenium		
Silver		
Thallium		
Zinc	Zinc	360

[a] In a particular plant actual concentration may vary from undetectable to ten times the typical value given.
[b] Variable to as much as 35 times the typical concentration given.

Traditional pollutant parameters

Average effluent quality for refineries with well operated treatment facilities (values in $mg\, l^{-1}$) would be:

BOD_5	5–15
COD	30–80
Total organic carbon	0.01–0.10
Total suspended solids	5–20
Ammonia	1–5
Sulphide	<0.10
Oil and grease	1–5

Treated effluent quality at a particular plant may vary from these average values, particularly if treatment systems are not operating properly.

Toxic pollutant parameters
Treated effluent quality is addressed here in terms of the pollutants listed in Table 1.6. Of the volatile organics listed, typically only benzene may be detected in the treated effluent. In plants where benzene is generated in the process wastewater, a typical concentration in the treated effluent is $2\,\mu g\,l^{-1}$. In a particular refinery, however, the actual benzene concentration may vary from being undetectable to several times greater than this typical value. Of the semi-volatile organics and pesticides listed, typically none is detected in the treated effluent. Of the metals listed, the following may be detected in the treated effluent at these typical concentrations:

Arsenic ($\mu g\,l^{-1}$)	10
Chromium ($\mu g\,l^{-1}$)	115
Copper ($\mu g\,l^{-1}$)	25
Cyanide ($\mu g\,l^{-1}$)	40
Lead ($\mu g\,l^{-1}$)	15
Nickel ($\mu g\,l^{-1}$)	5
Zinc ($\mu g\,l^{-1}$)	200

In a particular plant, however, actual concentration of these metals may vary considerably. Copper is more variable than the other metals listed above.

1.3 Discharge standards

1.3.1 Introduction
It is often said that 'necessity is the mother of invention'. In many respects, this is applicable to pollution control. Discharge regulations and standards create a need for treatment technology, which is ultimately satisfied through research, development and full-scale application.

In the control and treatment of wastewaters from the petroleum refining industry, a wide variety of technology is employed. In some instances, treatment techniques have been developed specifically for petroleum refining wastewaters. More commonly, however, methodologies utilized on wastewaters from other industries and processes used on domestic sewage have been applied to this specific need. Although the impact of discharge regulations and standards on the development of control technology in the petroleum refining industry is to some extent a matter of conjecture, it is apparent that a strong relationship exists.

An examination of the application in petroleum refineries of the specific in-plant controls and end-of-pipe treatment processes described in the next section reveals a distinct historical pattern. During the early period of development in the refining industry, concern for water pollution abatement was focused primarily on visible pollutants such as oil and grease and suspended solids. This emphasis led to the development and widespread usage of API separators for removal of these pollutants by gravity separation. As discharge standards for oil and grease and suspended solids were promulgated and ultimately became more and more stringent, other more sophisticated treatment processes such as dissolved air flotation and filtration were employed.

Similarly, with the advent of regulations on BOD, phenol and TOC, the use of secondary treatment in petroleum refineries has become widespread. This has generally been accomplished through the use of biological treatment processes such as activated sludge, aerated lagoons and rotating biological contactors, but has also included physicochemical treatment processes in a few instances. With the current regulatory focus on toxic pollutant control, particularly in the USA, emphasis has shifted to control techniques such as recycle and reuse, and treatment processes such as activated carbon adsorption. Thus, in a general way, control and treatment technology for water pollution abatement in the petroleum refining industry has evolved over the years essentially in response to the impetus of discharge regulations and standards.

1.3.2 Petroleum refinery discharge standards

Discharge standards for petroleum refineries are of course directly linked to the wastewater regulations and water pollution control policies in effect within the country concerned. In general, increasing public attention to environmental quality and the growing pressures of population and industry on limited environmental resources have led to environmental protection programmes in most countries. Nevertheless, discharge regulations and environmental policies vary widely, depending upon a great number of factors which include the many economic, social, and political considerations and goals which affect national policies with regard to pollution control and environmental quality.

Typically, the developing countries are not sympathetic to the view that economic growth should be curtailed to prevent environmental degradation: rather, they view industrialization and development as more important than protection of the environment. As a result, although more than half of the developing nations have passed environmental laws, enforcement is generally weak (Anon., 1978). Conversely, most industrialized nations have suffered the effects of pollution and associated environmental degradation in the past. This experience has generally led to strict pollution control legislation and regulations and rigorous enforcement.

This generalized scenario of world pollution control policies is significant in the context of this discussion, in that petroleum refineries are primarily concentrated in the industrialized nations. Consequently, a review of discharge standards in Western Europe, Japan, and the USA covers approximately two-thirds of the operating petroleum refineries in the world.

Western Europe

General public concern with environmental quality in Western Europe dates from as early as 1951 with the passage of the Rivers (Prevention of Pollution) Act in the UK, and 1957 with the passage of water pollution control legislation in West Germany. In the late 1960s and early 1970s wide-scale public reaction to environmental degradation led to general and specialized environmental laws (Table 1.7) and the development of institutions to implement them (OECD,

Table 1.7 Major water pollution control legislation in Western Europe (OECD, 1979)

Country	Year(s) of passage
Austria	1959
Belgium	1971
Finland	1961, 1965
France	1964
West Germany	1957, 1976
Greece	1978
Republic of Ireland	1977
Italy	1976
Netherlands	1970, 1975
Norway	1970
Sweden	1969
Switzerland	1971
United Kingdom	1951, 1961, 1974

1979). In 1975, the Commission of the European Communities identified more than 20 000 associations for the protection of the environment in Europe, and considered the estimate to be low.

In some countries, the environmental responsibilities of existing ministries were enlarged to handle the increased activity. This is the case, for instance, with the Ministry of Agriculture in Sweden, the Ministry of Health in the Netherlands, and the Ministries of the Interior in West Germany and Finland. In other countries, such as the UK and, since 1978, France, larger ministries, also responsible for housing and urban affairs, were formed to address environmental concerns. In yet other countries, such as Norway, more specialized Environmental Ministries were created (OECD, 1979).

Practically all countries in the EEC have set discharge standards, and the responsibility for doing so can rest with either central or local governments. In some countries, such as the UK and France, standards are formally or informally negotiated for sectors of industry, for specific firms or even for individual plants. In addition to discharge regulations, several countries, including Finland, France, West Germany, the Netherlands and the UK, have instituted pollution charges and taxes. Typically though, pollution charges are not widely used even by those countries that have adopted them. In the UK, charges are only applicable to discharges to sewers. In general, discharge regulations are more commonly relied upon to control water pollution (OECD, 1979).

Japan

Environmental awareness and enactment of pollution control laws in Japan has closely paralleled the pattern in the rest of the industrialized world. In the late 1960s, several severe environmental incidents, highlighted by the mass media, led to enactment of the Basic Law for Environmental Pollution Control in 1967 (Anon., 1979). Subsequently, under intense public pressure, the Japanese Diet passed a wave of additional strict environmental legislation. Pollution control, thenceforth, became one of the government's top priorities.

The legislation was supported by sweeping court decisions which required polluters to compensate the victims of the most serious pollution incidents: the residents of Minamata and Niigata paralyzed and killed by mercury poisoning in their fish; the sufferers of itai-itai ('ouch-ouch') disease in Toyama thought to be caused by cadmium poisoning; and the asthma, bronchitis, and emphysema cases in Yokkaichi with its mammoth petrochemical complexes (Meyerson, 1980). The courts ruled that enterprises must monitor all the risks of their pollution and adopt the best control technology available in the world. This strict interpretation of the law has also been combined with rigorous enforcement, but in addition strong financial incentives, including rapid depreciation and low interest loans, have been available to Japanese industry to promote compliance.

In terms of discharge regulations, the Water Pollution Control Law provides for the setting of uniform national effluent standards for industries with facilities that discharge into public waters. By 1979, 560 industrial categories, including petroleum refining, were subject to such effluent standards. In addition, the law provides that stricter effluent standards may be set by prefectural ordinances where it is judged that the national standards are insufficient to attain the water quality designated. At present, all of the prefectures have put such stricter standards into force. The waste load allocation concept has also been employed in specific areas, most notably the Seto Inland Sea, effectively to reduce the total pollution load being discharged in the area. All of these measures have had a marked effect in reducing water pollution and improving the overall environment in Japan (Anon., 1979).

USA

The Federal Water Pollution Control Act Amendments of 1972 established a comprehensive programme in the USA to 'restore and maintain the chemical, physical, and biological integrity of the Nation's waters'. The Amendments represented the culmination of the federal water pollution abatement effort in the USA, begun by Congress in 1948 and reinforced in 1956 by passage of the Federal Water Pollution Control Act, with subsequent amendments in 1961, 1965, 1966 and 1970.

The 1972 Amendments were actually a radical departure from the previous measures in that they made a number of fundamental changes in the approach to achieving clean water. One of the most significant changes was from an emphasis on the ambient quality of streams to direct control of effluents through the establishment of guidelines and standards which form a basis for the issuance of discharge permits. The Act required the development of industry-wide uniform national effluent limitations guidelines, based on the availability of technology to remove specific pollutants from discharges.

The guidelines were structured as three separate levels of effluent limitations: those required of existing industrial dischargers by 1977; more stringent limitations on existing dischargers required to be met by 1983; and the most restrictive standards applicable to all new sources within the industrial category. Effluent regulations were promulgated for the petroleum refining industry in 1974, setting mass limits, based on refinery throughput, on the discharge of BOD, COD, total suspended solids, oil and grease, phenolic compounds, ammonia, sulphide, total chromium, hexavalent chromium and pH (EPA, 1974a). The regulations were based on a mathematical flow model of the industry, which had been subdivided into five subcategories for regulatory purposes.

A court challenge by the petroleum industry in 1976 resulted in the invalidation of the 1983 standards. At the same time, a number of environmental action groups brought suit against the US Environmental Protection Agency (EPA), contending that the Agency had not been responsive to the requirements of the law regarding standards for controlling the discharge of toxic pollutants. This led to a settlement agreement in 1976 and revision of the law in 1977, directing EPA to develop discharge standards for 65 specific groups of toxic pollutants in 21 industrial categories including petroleum refining.

Extensive field sampling of refineries in the USA was conducted by EPA, the results generally indicating that toxic pollutants are not a major problem in petroleum refinery wastewaters after treatment. Accordingly, EPA proposed regulations in December 1979, on just two of the 65 toxic pollutants, namely chromium (total and hexavalent) and phenol. More stringent controls for existing refineries were also proposed for BOD, total suspended solids and oil and grease, based on a wastewater flow reduction of approximately 50% in the average refinery. These standards were intended to replace those invalidated by the courts in 1976.

1.4 Control and treatment technology

1.4.1 In-plant control
In-plant source control affords two major benefits:

(1) the overall reduction of pollutant load that must be treated by an end-of-pipe system;

(2) the reduction or elimination of a particular pollutant parameter before dilution in the main wastewater stream.

Proper in-plant control encompasses housekeeping to reduce wastewater volume and pollutant loading; segregation to insure effective wastewater management; and process modifications and recycle and reuse schemes to reduce the combined plant effluent requiring treatment.

Housekeeping

A significant degree of waste load reduction can be achieved in petroleum refineries by ensuring that effective housekeeping measures are practised. Among the procedures which can be used to minimize waste loading are the following:

- Efficient product sampling techniques designed to minimize the volume of product lost to waste during sampling.
- Efficient spill cleanup methods such as dry cleanup procedures, and well trained and organized crews to ensure rapid cleanup, to prevent spills from spreading.
- The use of vacuum truck equipment to collect spilt material for return to slop oil storage.
- Effective equipment maintenance to minimize leakage.
- Provision of curbing around process areas to contain spills and segregate oily stormwater.
- Periodic sewer maintenance, including flushing to prevent shock loads at the treatment plant as a result of sludge deposit washouts.
- Holding of washdown wastes during maintenance shutdowns, to allow gradual metering into treatment systems.
- Reducing excess water usage by shutting down pump gland cooling-water lines on pumps that are out of service, and shutting off washdown hoses when not in use.

Segregation

Effective management and waste load reduction in petroleum refineries requires that wastewaters be segregated adequately to ensure that they can receive the most effective treatment and/or reuse. In effect, all in-plant treatment options require the segregation of the process waste streams under consideration. If there are multiple sources of the particular pollutant or pollutants, they all require segregation from the main wastewater sewer. However, similar sources can be combined for treatment in one system. Sour waters constitute a type of wastewater that is produced at various locations within a refinery complex, but that can be treated as one combined wastewater stream.

A typical segregation scheme for a petroleum refinery would provide a 'clean' water sewer, an oily water sewer and a high contamination sewer (Beychok,

1967). Provision of such a segregated system allows the oil separation units to be sized to handle only those flows containing significant oil, while bypassing oilfree waters. Likewise, the final treatment process may be designed only for the high contamination and oily water streams, while the high-solids clean water stream may only require settling before discharge or reuse.

The major contributor to the high-solids clean water stream is stormwater, if once-through cooling is not practised. Stormwater from oily areas can be highly contaminated, and thus should be discharged to the oily water system. A stormwater surge pond should also be provided to prevent high-intensity storm runoff from overloading treatment facilities. A container for contaminated stormwater allows settling of solids which can otherwise cause emulsions to form. The surge capacity allows flow to be metered into the treatment facility gradually, to allow efficient treatment. Storm retention ponds should typically be designed on the basis of the maximum 10 year, 24 hour storm. A bypass with skimming facilities should be provided to divert only the tail end of larger storms which will carry a lower contaminant load than the leading edge of the storm.

Process modifications. Many new and modified refineries incorporate reduced water use and pollutant loading into their design. These modifications include:

- Replacement of barometric condensers on vacuum distillation units with surface (non-contact) condensers. This modification will reduce a major source of emulsified oily water. Alternatively, the oily cooling-water produced by barometric condensers may be recirculated in its own oily water cooling system with only periodic blowdown.
- Replacement of steam jet ejectors by vacuum pumps on vacuum distillation units. This will eliminate the oily steam condensate produced by the former.
- Substitution of air cooling for some cooling requirements, thereby reducing cooling-water blowdown.
- Replacement of steam stripping columns with reboiler units for the stripping of lighter ends from draw-offs. This change will reduce the production of foul condensates by reducing the use of direct contact steam.
- Substitution of improved catalysts with higher activity and longer life, which require less regeneration and therefore produce a lower waste load.
- Use of hydrocracking and hydrotreating processes which produce lower wastewater loadings than the older processes.
- Increased use of improved drying, sweetening and finishing procedures to minimize spent caustics and acids, water washes and filter solids requiring disposal.
- Recycle of wastewater at the process units to reduce the amount of wastewater leaving the process area.

Recycle and reuse
There are several possibilities, within typical petroleum refinery processes, for

reduction of wastewater discharge by recycling and reusing various wastewaters elsewhere in the refinery.

Wastewaters emanating from end-of-pipe facilities are generally of such quality that reuse can be quite attractive. Uses for treated refinery wastewaters include makeup to cooling towers, pump gland cooling systems, washdown waters and fire-water systems.

Cooling-water systems. Finelt and Crump (1977) report that refiners faced with increasing fresh water costs may direct their water management policies toward the recirculation of treated water. Properly treated wastewater can be recycled as makeup to the cooling tower system. At new refineries, it is possible that the recycle system could be justified economically over a non-recycle system. There are a number of factors to be considered, most notably the cost of fresh water, in determining the least costly system. At existing well-operated facilities, very high fresh water costs are required before a recycle system proves to be less costly than a non-recycle system. However, application of recycle technology can reduce effluent discharge by up to 90% and possibly help towards the goal of 'zero discharge' facilities.

In many circumstances, stripped sour waters can be used as makeup to a recirculating cooling system. The cooling tower of the system provides effective phenol oxidation. However, care must be exercised with this option because problems may be encountered with corrosion and fouling in condensers due to slime growth promoted by biological activity.

A similar proposition is the use of well treated municipal effluent for cooling-water makeup. This generally does not cause corrosion problems, but slime growth potential is still high. In general, some slimicide control is essential with this reuse option.

Reed *et al.* (1977) report that side stream softening can greatly reduce cooling tower blowdown. They conclude that bypass lime softening appears to be a realistic technique applicable to both new and existing cooling tower systems. Cooling-water systems in high dissolved solids applications can be expected to be adequately protected through the use of corrosion inhibitors. Reed *et al.* present a method for designing for zero blowdown application and for predicting steady state conditions. A pilot cooling tower was operated with a recirculating water chemistry designed to simulate the ultra-high TDS levels representative of conditions expected for zero blowdown systems. Low corrosion rates on mild steel and copper alloy specimens were reported after application of a proprietary synergistic chromate corrosion inhibitor. These low corrosion rates were especially pertinent to the application of zero blowdown systems at existing installations which have a considerable amount of mild steel and copper plumbing and process equipment.

Crude desalter makeup. The use of sour waters as makeup to the crude desalter is a proven technology in this industry. This practice removes approximately

80–90% of the phenol present, because the phenolics are extracted from the sour water into the crude being washed. However, the removal efficiency varies greatly depending on a number of factors, and this treatment scheme may not be a practical alternative for some refineries (EPA, 1976).

Certain crudes, particularly California crudes, may present problems in reusing sour waters in the desalter because they produce emulsions in the desalter effluent. Also, this use requires close control because excessive ammonia or sulphide concentrations in the desalter feed can cause excessive corrosion in the desalter or in the overhead accumulator of the atmospheric distillation unit. Moreover, excessive phenol contribution from the feed can apparently produce an off-colour kerosene cut.

Normally, vacuum condensates may be reused directly in the desalter, without need for stripping. However, the combined volume of stripped condensate and vacuum condensate usually exceeds desalter feed requirements, necessitating alternatives for disposal.

Close control must be maintained on desalters, because upsets can result in the drawoff of emulsified oils which will upset treatment facilities by shock loading. At the other extreme, carryover of water, silt and salts to the atmospheric tower can cause corrosion, line and pump plugging, and generally result in hazardous operating conditions.

Clean condensate recycle. Generally, refineries have been designed to provide for several progressive uses of non-contact steam, from higher to lower pressure, in order to maximize energy conservation. This is certainly being maximized in newer refineries, with increased consciousness of energy conservation needs. However, the final clean condensate is often discharged to sewers at many small scattered outlets throughout the plant. It is desirable to collect the maximum quantity of these clean condensates and return them to boiler feed, in order to reduce boiler feed-water intake and treatment requirements. However, difficulties may be encountered in collecting many small discharges of condensate during extremely cold weather.

Clean stormwater reuse. Clean or lightly contaminated stormwater may be collected and used as a raw water source. In particular, the low hardness of stormwater makes it a desirable source for cooling-water makeup or related applications, where scaling may be a problem.

As an example, it is estimated that a $15\,900\,m^3$ refinery using 65% recirculated water cooling and 35% air cooling located in an area with annual rainfall of 75 cm could supply its complete cooling-water makeup requirements from collected stormwater (Fern, 1974). In such cases, it is necessary to consider the precipitation pattern and the degree of stormwater contamination, along with the treatment requirements necessary to ensure consistent quality feed.

Pump cooling-waters. Another method of wastewater reduction is the elimination of cooling-water from general-purpose pumps. In certain instances the elimination of water can increase machinery reliability, reduce capital expenditure on piping and water treatment facilities and save operating costs. Guidelines are available for implementation of a well planned step-by-step programme of deleting cooling-water from pumps and drivers. These procedures have been successfully implemented on a full-scale basis.

A typical refinery may have 70 pumps each demanding daily 11–27 m^3 water for cooling purposes (Fern, 1974). This cooling demand can be met by a separate recirculating water–glycol cooling system, analogous to the cooling system in an automobile. Such a closed-loop system would prevent evaporative losses of cooling-water resulting in dissolved solids buildup and consequent blowdown requirements.

In addition to the specific recycle schemes outlined above, many refinery wastewater streams, such as treated sour waters, cooling tower blowdowns and utility blowdowns, are suitable for use as wash-waters and fire system water. However, reuse of wastewaters for these purposes requires investigation on a plant-by-plant basis to determine the technical and economic feasibility.

In general, it is recommended that reuse be approached cautiously because of the sensitivity of many processes to upsets caused by recycle schemes. Also, it should be emphasized that reduction in water usage sometimes may be more cost effective in reducing the quantity of wastewater discharge than water reuse or recycle.

1.4.2 In-plant treatment
A number of in-plant treatment schemes are being implemented or are available for use in petroleum refineries involving specific waste streams.

Sour waters
Sour waters generally result from water brought into direct contact with a hydrocarbon stream. This occurs when steam is used as a stripping or mixing medium, or when water is used as a washing medium, as in the crude desalting unit. Sour waters contain sulphides, ammonia and phenols.

The most common in-plant treatment schemes for sour waters involve sour water stripping, sour water oxidizing or combinations of the two. Sour water stripping is a gas–liquid separation process that uses steam or flue gas to extract the gases (sulphides and ammonia) from the wastewater. The stripper itself is a distillation-type column containing either trays or packing material. Columns range from simple one-pass systems to sophisticated reflux columns with reboilers. Some refineries have a number of units operating in parallel, while others use two columns in series to facilitate high ammonia removals (Chevron WWT process). The vast majority of units used in the USA utilize steam as the stripping medium.

In removing sulphides and ammonia, the efficiency of sour water treatment processes is greatly influenced by pH. In general, sour water strippers remove between 85 and 99% of the sulphides present. However, when the pH is lowered by means of acid treatment, stripping efficiency is increased. On the other hand, when caustic is utilized to maintain high pH, up to 95% ammonia removal can be achieved. By considering pH in the stripping process, one can either adjust the pH to optimize removal of both sulphides and ammonia, or use a two-stage sour water stripping process to obtain maximum removal of both pollutants.

Existing sour water stripper performance can be improved by:

- increasing the number of trays;
- increasing the steam rate;
- increasing tower height;
- adding a second column in series.

Another way of treating sour water is to oxidize by aeration. Compressed air is injected into the waste, followed by sufficient steam to raise the reaction temperature to at least 88°C. Reaction pressure of 350–700 kPa is required. Oxidation proceeds rapidly and converts practically all the sulphides to thiosulphates and about 10% of the thiosulphates to sulphates. Air oxidation, however, is much less effective than stripping in regard to reduction of the oxygen demand of sour waters, since the remaining thiosulphates can later be oxidized to sulphates by aquatic micro-organisms. Oxidation systems using peroxide and chlorine have also been identified. These systems operate in open tanks, without the use of steam.

Because of the very low limits required by the County Sanitation Districts of Los Angeles County, refineries discharging to this sewer system use both sour water strippers and sour water oxidizers in series. Levels of less than $0.1\,\text{mg}\,\text{l}^{-1}$ sulphides in the effluent are consistently maintained by these refineries. Los Angeles County also maintains a restriction of $50\,\text{mg}\,\text{l}^{-1}$ of thiosulphates to control the chlorine demand at the sewage treatment plant.

Spent caustic
Spent caustic solutions are generated by various finishing wet treatment processes aimed at neutralizing and extracting acidic materials occurring naturally in crude acidic products from various chemical treatment steps, and acidic materials produced in cracking processes. Spent caustics generally contain sulphides, mercaptans, sulphates, sulphonates, phenols and naphthenic acids. The phenol concentrations, in particular, may be high enough to warrant processing of spent caustics for the recovery of phenols.

It is common practice in many refineries to neutralize spent caustic with spent acid from sulphuric acid alkylation and charge the neutralized solution to a sour water stripper. If the bottoms are then sent to crude desalting, the high phenol content may be recovered within the process by extraction back into the crude feed. Where spent acid from sulphuric acid alkylation is not available, spent caustics may be neutralized with flue gas.

In some circumstances, spent caustics may be oxidized to convert sulphides to thiosulphates. Typically 85–99% sulphide conversion can be achieved in this way. However, this process is not effective for caustics with high phenol content because phenols tend to inhibit oxidation reactions.

Slop oil treatment

Product streams that fail to meet production specifications, product and crude spills and oil recovered from wastewater streams all contribute to form slop oil, which can be reprocessed to recover usable hydrocarbons. However, it is necessary to treat slop oils to prevent them from causing upsets to normal crude processing.

One treatment used is heating to 90°C for 12–14 hours to allow separation of water content. Normally, a clean oil layer results, above a middle layer of oil–water emulsion and a bottom layer of water, solids and some oil. It may sometimes be necessary to add de-emulsifiers to break slop oil emulsions. In addition, treatment by centrifugation or precoat filtration using diatomaceous earth filters may be required.

Slop oil recovery and treatment should be accompanied by adequate storage to allow gradual feeding of recovered slop oil to the process to avoid unnecessary upsets.

Cooling tower blowdown

Removal of metals (such as chromium and zinc) and phosphate can be achieved by precipitation and clarification at a relatively high pH (8–10). Hexavalent chromium, however, must be first reduced to the trivalent state before it can be precipitated and removed by clarification. This usually is accomplished by the addition of sulphur dioxide, ferrous sulphate or sodium bisulphite. The pH of the wastewater then rises with the addition of lime or caustic (lime is preferred if phosphates are to be precipitated); clarification of the wastewater stream follows. Flocculants and flocculant aids, such as ferric chloride, alum and polymers, can be added to increase removal efficiencies.

EPA (1974b) reports that the following effluent concentrations ($mg\,l^{-1}$) can be obtained from these types of metal precipitation systems:

Chromium, total	0.3
Chromium, hexavalent	0.06
Zinc	0.3

Solvent extraction

Liquid–liquid solvent extraction is the separation of the constituents of a liquid solution by contact with another immiscible liquid. If the substances comprising the original solution distribute themselves differently between the two liquid phases, a certain degree of separation results, and this may be enhanced by use of multiple contacts.

Solvent extraction separates the original solution into two streams: a treated stream (the raffinate) and a recovered solute stream (which may contain small amounts of water and solvent). Solvent extraction may thus be considered a recovery process since the solute chemicals are generally recovered for reuse, resale or further treatment and disposal.

A process for solvent extracting a solution typically includes three basic steps: the actual extraction, solute removal from the extracting solvent and solvent recovery from the raffinate (treated stream). The process may be operated continuously.

The first step, extraction, brings the two liquid phases (feed and solvent) into intimate contact to allow solute transfer either by forced mixing or by countercurrent flow caused by density differences. The extractor also has provision to allow separation of the two phases after mixing. One output stream from the extractor is the solute-laden solvent; some water may also be present. Solute removal may be via a second solvent extraction step, distillation or some other process. Solvent recovery from the treated stream may be required if solvent losses would otherwise add significantly to the cost of the process or cause a problem with the discharge of the raffinate. Solvent recovery may be accomplished by stripping, distillation, adsorption or other suitable process.

Removals of up to 99% of phenols present in refineries have been achieved using solvent extraction (EPA, 1980).

1.4.3 End-of-pipe control technology
End-of-pipe control technology in the petroleum refining industry relies heavily upon the use of biological treatment methods. These are supplemented by appropriate pretreatment to insure that proper conditions, especially sufficient oil removal and pH adjustment, are present in the feed to the biological system. When used, initial treatment most often consists of neutralization for control of pH or provision of equalization basins to minimize shock loads on the biological system. The incorporation of solids removal before biological treatment is not as important as it is in treating municipal wastewater.

Preliminary treatment
Equalization. The purpose of equalization is to dampen out surges in flows and loadings. This is especially necessary for a biological treatment plant, as high concentrations of certain materials will upset or completely inactivate the bacteria in the treatment plant. By evening out the loading on a treatment plant, the equalization step enables the treatment plant to operate more effectively and with fewer maintenance problems. Without equalization, an accident or spill within the refinery can greatly affect the effluent quality or kill the biomass of a biological treatment process.

The equalization step usually consists of a large pond that may contain mixers to provide better dampening of waste fluctuations. In some refineries the equalization is done in a tank. Equalization can precede or follow gravity separation, but is more effectively carried out first as it increases the overall efficiency of the separator. However, care must be taken to prevent anaerobic decomposition in the equalization facilities.

Gravity separation of oil. Gravity separators remove most of the free oil found in refinery wastewaters. Because of the large amounts of reprocessable oils that can be recovered in gravity separators, these units are considered to be an integral part of the refinery processing operation rather than a wastewater treatment process. The functioning of gravity-type separators depends upon the difference in specific gravity of oil and water. The gravity-type separator will not separate substances in solution, nor will it break emulsions. The effectiveness of a separator depends upon the temperature of the water, the density and size of the oil globules and the amount and characteristics of the suspended matter present in the wastewater.

The API separator is the most widely used gravity separator. The basic design is a long rectangular basin, with enough detention time for most of the oil to float to the surface and be removed. Most API separators are divided into more than one bay to maintain laminar flow within the separator, making the separator more effective. API separators are usually equipped with scrapers to move the oil to the downstream end of the separator, where the oil is collected in a slotted pipe or on a drum. On their return to the upstream end, the scrapers travel along the bottom moving sludge solids to a collection trough. Any sludge which settles can be dewatered and either incinerated or disposed of as landfill.

A schematic of a typical API separator is shown in Fig. 1.9. The correct design of API separators depends upon the following parameters:

- Specific gravity of the oil phase
- Diameter of the oil globules
- Operating temperature
- Influent flow
- Oil content
- Suspended solids content
- Percentage of influent pollutants susceptible to gravity separation

However, the role of the API separator in the refinery wastewater treatment system must be considered in relation to its limitations, namely:

- It can only remove gravity-separable oils and will not separate stable emulsions.
- It provides limited reduction of other parameters.
- It must be designed to minimize short-circuiting.
- The presence of any kind of emulsifier, spent caustic, detergent, etc., will upset efficiency.

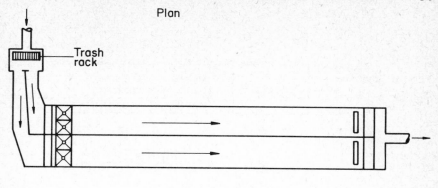
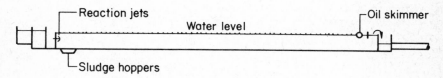

Fig. 1.9 API oil–water separator
Typical dimensions: depth 0.9–2.4 m; width 1.8–6.0 m; depth/width ratio 0.3–0.5

- Suspended solids attached to oils may not settle.
- Hydraulic overloading reduces efficiency.
- If provision is not made for continuous sludge removal, two full-size basins are required, to allow periodic sludge removal.

Properly designed and operated separators normally produce effluents with oil concentrations of less than 50 mg l^{-1}.

Parallel plate separator. The parallel plate separator is a device which is very similar to the API separator. It was developed to improve oil and solids removal by mounting parallel plates approximately 15 cm apart at 45° to the vertical along the length of the separator. By vastly increasing surface and floor area, this device permits more efficient collection of oil and solids.

Dissolved air flotation. Dissolved air flotation consists of saturating a portion of the wastewater feed, or a portion of the feed or recycled effluent from the flotation unit, with air at a pressure of 275–415 kPa. The wastewater or effluent recycle is held at this pressure for 1 to 5 minutes in a retention tank and then released at atmospheric pressure to the flotation chamber. The sudden reduction in pressure results in the release of microscopic air bubbles which attach themselves to oil and suspended particles in the wastewater in the

40 INDUSTRIAL WASTEWATER TREATMENT

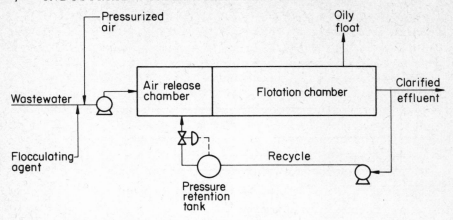

Fig. 1.10 Typical air flotation–flocculation system
Design criteria: Pressure (kPa) 170–480
Ratio air/solids 0.01–0.1
Detention time (min) 20–60
Surface hydraulic
  loading ($m^3\ m^{-2}\ d^{-1}$) 20–330
Recycle, where employed (%) 5–120
Solids loading
  ($kg\ m^{-2}\ h^{-1}$) 2.4–24.0

flotation chamber. This results in agglomerates which, because of the entrained air, have greatly increased vertical rise rates of about 0.0025 to 0.005 m s^{-1}. The floated materials rise to the surface to form a froth layer. Specially designed flight scrapers or other skimming devices continuously remove the froth. The retention time in the flotation chambers is usually about 10 to 30 minutes. The effectiveness of dissolved air flotation depends upon the attachment of bubbles to the suspended oil and other particles which are to be removed from the waste stream. The attraction between the air bubble and particle is a result of the particle surface and bubble-size distribution.

Chemical flocculating agents, such as salts of iron and aluminium, with or without organic polyelectrolytes, are often helpful in improving the effectiveness of the air flotation process and in obtaining a high degree of clarification.

Dissolved air flotation is used by a number of refineries to treat the effluent from the oil separator. Dissolved air flotation using flocculating agents is also used to treat oil emulsions. Specific demulsifying chemicals are used in this process, depending on the zeta potential of the emulsion particles. The use of appropriate demulsifying chemicals can provide a cost-effective and highly efficient dissolved air flotation process. When this process is complete, the froth is skimmed from the flotation tank and can be combined with other sludges (such as those from a gravity separator) for disposal. Figure 1.10 shows a typical dissolved air flotation system (AWWA, 1971).

Biological treatment

Oxidation ponds. The oxidation pond is practical where land is plentiful and cheap. An oxidation pond has a large surface area and a shallow depth, usually not exceeding 2 m. These ponds have long detention periods of 11–110 days.

The shallow depth allows the oxidation pond to be operated aerobically without mechanical aerators. The algae in the pond produce oxygen by means of photosynthesis. This oxygen is then used by the bacteria to oxidize the wastes. Because of the low loadings, little biological sludge is produced and the pond is fairly resistant to upsets due to shock loadings.

Oxidation ponds are usually used as the major treatment process. However, some refineries use ponds as a polishing process after other treatment processes. This process is not reliable in very cold climates.

Aerated lagoons. The aerated lagoon is a smaller, deeper oxidation pond equipped with mechanical aerators or diffused air units. The addition of oxygen enables the aerated lagoon to have a higher concentration of microbes than the oxidation pond. The retention time in aerated lagoons is usually shorter, between 3 and 10 days. Most aerated lagoons are operated without final clarification. As a result, biomass is discharged in the effluent, causing the effluent to have high BOD_5 and solids concentrations. Where effluent standards are stringent, final clarification is necessary. The effectiveness of conventional clarification on such effluents is often poor. Filtration may be necessary to satisfy suspended solids limitations.

Trickling filters. A trickling filter is an aerobic biological process. It differs from other processes in that the biomass is attached to the bed medium, which may be rock, slag or plastic. The filter works by:

(1) adsorption of organics by the biological slime;
(2) diffusion of air into the biomass;
(3) oxidation of the dissolved organics.

When the biomass reaches a certain thickness, part of it sloughs off. When the filter is used as the major treatment process, a clarifier is used to remove the sloughed biomass.

The trickling filter can be used either as the complete treatment system or as a roughing filter. Most applications in the petroleum industry use it as a roughing device to reduce the loading on an activated sludge system.

Typical high-rate trickling filter performance and design criteria are summarized as follows (EPA, 1978):

Performance. Single-stage configuration with filter effluent recirculation and primary and secondary clarification (per cent removal):

BOD_5	60–80%	Phosphorus	10–30%
NH_4-N	20–30%	SS	60–80%

Design criteria:

Hydraulic loading (with recirculation)	$9.4-47 \, m^3 \, m^{-2} \, d^{-1}$
Recirculation ratio	0.5:1 to 4:1
Dosing interval	$\leq 15 \, s$ (continuous)
Sloughing	Continuous
Medium	Rock, 2.5–12.7 cm (using square mesh screen)
Organic loading	$0.32-0.96 \, kg \, BOD_5 \, m^{-3} \, d^{-1}$
Bed depth	0.9–1.8 m
Underdrain minimum slope	1%

Rotating biological contactors (RBC). RBCs are analogous to trickling filters, in that they are fixed-film reactors. Bacterial slime grows on plastic discs rotating through the wastewater. Approximately half of the circular disc is out of the water at any one time, being aerated, and half is under water supporting biological growth. The discs rotate slowly, requiring minimal power. Discharge from an RBC may need clarification to remove the suspended material generated in the fixed growth process. Although this process is a relatively new development, it has been successfully applied in several cases to petroleum refinery wastewater.

Typical performance and design criteria for RBCs are summarized as follows (EPA, 1978):

Performance. Four-stage system with final clarifier and preceded by primary treatment (per cent removal):

BOD_5	80–90%	SS	80–90%
Phosphorus	10–30%	NH_4-N	95%*

*Dependent upon temperature, alkalinity, organic loading and unoxidized nitrogen loading.

Design criteria:

Organic loading:	
without nitrification	$0.48-0.96 \, kg \, BOD_5 \, m^{-3} \, d^{-1}$
with nitrification	$0.24-0.32 \, kg \, BOD_5 \, m^{-3} \, d^{-1}$
Hydraulic loading:	
without nitrification	$0.03-0.06 \, m^3 \, m^{-2} \, d^{-1}$ of media surface area
with nitrification	$0.012-0.024 \, m^3 \, m^{-2} \, d^{-1}$ of media surface area
No. of stages per train	≥ 4
No. of parallel trains	Recommended ≥ 2

Rotational velocity (peripheral velocity)	30 cm s^{-1}
Per cent media submerged	40%
Tank volume	0.005 m^3 m^{-2} of disc area
Detention time based on 0.005 m^3 m^{-2}	
without nitrification	40–90 min
with nitrification	90–230 min
Secondary clarifier overflow rate	20–33 m^3 m^{-2} d^{-1}

Activated sludge. Activated sludge is an aerobic biological treatment process in which high concentrations (1500–3000 mg l^{-1}) of newly grown and recycled microbial biomass are suspended uniformly throughout a holding tank to which raw wastewaters are added. Oxygen is introduced by mechanical aerators, diffused air systems or a combination of the two. The organic materials in the waste are removed from the aqueous phase by the microbial biomass and stabilized by biochemical synthesis and oxidation reactions. The basic activated sludge process consists of an aeration tank followed by a sedimentation tank. The flocculant microbial growths removed in the sedimentation tank are recycled to the aeration tank to maintain a high concentration of active micro-organisms. Although the micro-organisms remove almost all of the organic matter from the waste being treated, much of the converted organic matter remains in the system in the form of microbial cells. These cells have a relatively high rate of oxygen demand and must be removed from the treated wastewater before discharge. Thus, final sedimentation and recirculation of biological solids are important elements of an activated sludge system.

Sludge is wasted on a continuous basis at a relatively low rate to prevent build-up of excess activated sludge in the aeration tank. Shock organic loads usually result in an overloaded system and poor sludge settling characteristics. Effective performance of activated sludge facilities requires pretreatment to remove or substantially reduce oil, sulphides and phenol concentrations. The pretreatment units most commonly used are:

- gravity separators and air flotation units to remove oil;
- sour water strippers to remove sulphides, mercaptans and phenol.

Equalization also appears to be necessary to prevent shock loadings from upsetting the aeration basin. Because of the high rate and degree of organic stabilization possible with activated sludge, application of this process to the treatment of refinery wastewaters has been increasing rapidly in recent years.

Many variations of the activated sludge process are currently in use. Examples include:

- the tapered aeration process, in which more air is added at the influent where the oxygen demand is the highest;

- step aeration, which introduces the influent wastewater along the length of the aeration tank;
- contact stabilization, in which the return sludge to the aeration tank is aerated for 1 to 5 hours.

The contact stabilization process is useful where the oxygen demand is in the suspended or colloidal form. The completely mixed activated sludge plant normally uses mechanical mixers to mix the influent with the contents of the aeration basin, decreasing the possibility of upsets due to shock loadings. The Pasveer oxidation ditch is a variation of the completely mixed activated sludge process that is widely used in Europe. Here brushes are used to provide aeration and mixing in a narrow oval ditch. The advantage of this process is that the mixed liquor suspended solids (MLSS) content can be higher than in the conventional activated sludge process because of the longer sludge age used, and the wasted sludge is relatively easy to dewater. At least one refinery uses the Pasveer ditch type system.

The activated sludge process has several disadvantages. Because of the amount of mechanical equipment involved, its operating and maintenance costs are higher than those of other biological systems. The small volume of the aeration basin makes the process more subject to upsets than either oxidation ponds or aerated lagoons.

The following presents typical performance and design data for activated sludge systems (EPA, 1978):

Performance:

BOD removal (conventional activated sludge system)	85–95%
Oxygen absorption efficiency (agitator sparger system under standard conditions in clean water)	10–20%
Oxygen absorption efficiency (diffusers in domestic wastewater activated sludge service)	3–15% (3–7% for coarse bubble 10–15% for fine bubble)
Oxygen transfer rate of agitator–sparger (standard conditions in clean water)	1.1–1.5 kg kW^{-1} h^{-1}
Oxygen transfer rate of mechanical surface-type aerators (standard conditions in clean water)	1.7–2.1 kg kW^{-1} h^{-1}

Generally, the power usage for a given oxygen transfer rate by an agitator–sparger system is 30–40% less than for coarse-bubble air diffusion systems.

Design criteria: a partial listing of design criteria for the conventional activated sludge process is summarized as follows:

Volumetric loading (kg $BOD_5\,m^{-3}\,d^{-1}$)	0.4–0.8
Aeration detention time (h) (based on average daily flow)	4–8
MLSS (mg l^{-1})	1500–3000
F/M (kg $BOD_5\,kg^{-1}$ MLSS d^{-1})	0.25–0.5
Air required ($m^3\,kg^{-1}$ BOD_5 removed)	50–94
Sludge retention time (d)	5–10

A partial listing of design criteria for the contact stabilization process is summarized as follows:

F/M (kg $BOD_5\,kg^{-1}$ MLSS d^{-1})	0.2–0.6
Volumetric loading (kg $BOD_5\,m^{-3}\,d^{-1}$)	0.5–0.8 (based on contact and stabilization volume)
MLSS (mg l^{-1})	1000–2500, contact tank 4000–10 000, stabilization tank
Aeration time (h)	
based on average daily flow	0.5–1.0, contact tank
based on sludge recycle flow	2–6, stabilization basin
Sludge retention time (d)	5–10
Recycle ratio, R	0.25–1.0
Std m^3 air kg^{-1} BOD_5 removed	50–131
kg $O_2\,kg^{-1}$ BOD_5 removed	0.7–1.0
Volatile fraction of MLSS	0.6–0.8

Granular media filtration

Granular media filtration utilizes a bed of granular particles to act as a filter in removing solids in water. There are several types of granular media filter: sand, dual and multimedia. Operation is basically the same, the only difference being the filter medium. The sand filter uses a relatively uniform grade of sand resting on a coarser material. The dual media filter has a coarse layer of coal above a fine layer of sand (Fig. 1.11).

The filter bed is supported by an underdrain system to remove the filtered water from the unit. As the water passes down through the filter, the suspended matter is trapped in the filter medium, thus reducing the filtration rate and increasing the pressure needed to force the water through the filter. When the pressure drop becomes excessive, the flow through the filter is reversed at a velocity sufficient to dislodge the collected solids from the filter medium. The backwash cycle occurs approximately once a day, depending on loading (EPA, 1978).

The following are typical performance and design data for granular media filters (EPA, 1978).

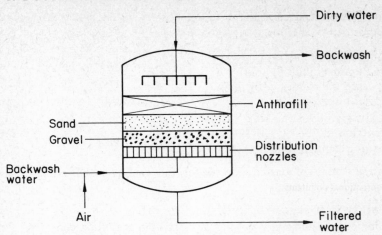

Fig. 1.11 High-rate dual media filtration

Performance:

Filter influent	Filter effluent BOD_5 (mg l^{-1})
High-rate trickling filter	10–20
Two-stage trickling filter	6–15
Contact stabilization	6–15
Conventional activated sludge	3–10
Extended aeration	1–5

Chemicals required. Alum, iron salts and polymers can be added as coagulant aids directly preceding filtration units. This, however, generally reduces the filter run length.

Design criteria:

Filtration rate ($m^3 m^{-2} min^{-1}$)	0.082–0.33
Bed depth (m)	0.6–1.2
(sand/anthracite depth ratio 1:1–4:1)	
Backwash rate ($m^3 m^{-2} min^{-1}$)	0.61–1.0
Filter run length (h)	8–48
Terminal head loss (m)	1.8–4.6

Granular activated carbon

Granular activated carbon has been used in the potable water industry for many years. More recently it has been used to remove dissolved organics in industrial and municipal wastewater treatment plants. Activated carbon systems have functioned both as polishing units following biological treatment systems and as the major treatment process in a physicochemical treatment system.

The granular activated carbon system considered here consists of one or more trains of carbon columns, each train having three columns operated in series. The columns operate by rotating their positions in the train, so that the newly regenerated carbon is in the third vessel, while the first vessel contains the most spent carbon.

Table 1.8 gives removal data for granular activated carbon adsorption (EPA, 1980).

Table 1.8 Pollutant removal by granular activated carbon adsorption (EPA, 1980)
Pilot scale study

Pollutant/parameter	Concentration		Removal
	Influent	Effluent	
	(mg l^{-1})	(mg l^{-1})	(%)
Conventional pollutants			
BOD$_5$	57	9	83
TOC	37	13	65
TSS	8	3	62
Oil and grease	12.3	1.8	85
Total phenol	2.7	0.02	99

Design/operating parameters
Unit configuration: upflow; 4 columns in series
Total flow: 0.002 m^3 min^{-1}
Hydraulic loading: 0.15 m^3 m^{-2} min^{-1}
Contact time: 36 min
Total carbon inventory: 0.071 m^3
Carbon type: Filtrasorb 300, 8 × 30 mesh

Powdered activated carbon

A new technology developed over the past several years consists of the addition of powdered activated carbon to biological treatment systems. The adsorbant quality of carbon, which has been known for many years, aids in the removal of soluble organic materials in the biological treatment unit. This treatment technique also enhances colour removal, clarification and system stability, as well as BOD and COD removal. Results of pilot testing indicate that this type of treatment, when used as a part of the activated sludge process, is a viable alternative to granular carbon systems.

Rizzo (1976) describes several tests using powdered activated carbon added to petroleum refinery activated sludge systems. Rizzo reports on a plant test of carbon addition to an extended aeration activated sludge unit at the Sun Oil Refinery in Corpus Christi, Texas. This test tried three carbon dosages: 24 mg l^{-1}, 19 mg l^{-1} and 9 mg l^{-1}. Test results showed that even the very small carbon dosages significantly improved BOD, COD and TSS removals, as well as producing uniform effluent quality, a clearer effluent and elimination of foam.

Grieves *et al.* (1977) report on a pilot plant study at the Amoco refinery in Texas City, Texas, where the addition of activated carbon to the activated sludge process was evaluated in 37.9 litre pilot plants. Significant improvements were observed in soluble organic carbon (53%), soluble COD (60%), NH$_3$-N

(98%) and phenolics after addition of 50 mg l^{-1} activated carbon. Improvement increased with increasing carbon dosage.

At higher carbon dosages, resulting in aerator levels of 1000 mg l^{-1} or more, Exxon got positive results. In a field test for which the scale was undisclosed, Thibault et al. (1977) significantly improved effluent quality and noted improvement in shock loading resistance, leading to process stability. Removals of TOC and COD increased on average by 10%.

Another powdered activated carbon scheme has been studied that uses very high sludge ages, or mean cell residence time (60 days or more). This allows the carbon to accumulate to high concentrations in the mixed liquor, even though only small makeup amounts are added to the system. This approach may eliminate the need for costly regeneration, since the low carbon addition rate allows the disposal of spent carbon with the sludge. Considerable pilot work has been done with this concept but no full-scale system is currently operating.

1.4.4 Summary of effluent treatment methods

As described above, the petroleum refining industry is using a wide range of treatment methods for the control of pollutant discharges. Table 1.9 presents a

Table 1.9 Summary of treatment technologies for 1973 and 1976

Treatment system	No. of user refineries in the USA	
	1973	1976
Corrugated plate separators	4	20
Chemical flocculation	1	46
Dissolved air flotation	56	68
Other flotation systems	1	15
Prefiltration	unknown	6
Activated sludge	30	50
Trickling filter	7	10
Aerated lagoon	63	73
Stabilization pond	44	35
Rotating biological contactor	0	5
Other organics removal	4	10
Filtration	10	23
Polishing ponds	unknown	75
Activated carbon	1	2
Evaporation or percolation ponds	26	37

summary of the use of these technologies in this industry in the USA. These data are based upon US EPA industry surveys in 1973 and 1976. As can be seen, the industry in the USA installed large numbers of treatment systems throughout the middle 1970s. This was primarily due to the need to meet state and national pollutant limitation requirements. Tables 1.10 and 1.11 summarize expected effluent concentrations and removal efficiencies for the treatment systems described in this section.

Table 1.10 Expected effluent quality from petroleum refinery wastewater treatment processes

Process	Effluent concentration (mg l^{-1})							
	BOD$_5$	COD	TOC	SS	Oil	Phenol	Ammonia	Sulphide
API separator	250–350	260–700	NA	50–200	20–100	6–100	15–150	NA
Clarifier	45–200	130–450	NA	25–60	5–35	10–40	NA	NA
Dissolved air flotation	45–200	130–450	NA	25–60	5–20	10–40	NA	NA
Oxidation pond	10–60	50–300	NA	20–100	1.6–50	0.01–12	3–50	0–20
Aerated lagoon	10–50	50–200	NA	10–80	5–20	0.1–25	4–25	0–0.2
Activated sludge	5–50	30–200	20–80	5–50	1–15	0.01–2.0	1–100	0–0.2
Trickling filter	25–50	80–350	NA	20–70	10–80	0.5–10	25–100	0.5–2
Granular media filter	NA	NA	25–61	3–20	3–17	0.35–10	NA	NA
Granular activated carbon	3–10	30–100	1–17	1–15	0.8–2.5	0–0.1	1–100	0–0.2
Powdered activated carbon	5–50	15–200	NA	NA	NA	NA	NA	NA

NA = Data not available.

Table 1.11 Typical removal efficiencies for petroleum refinery wastewater treatment processes

Process	Removal efficiency (%)							
	BOD$_5$	COD	TOC	SS	Oil	Phenol	Ammonia	Sulphide
API separator	5–40	5–30	NA	10–50	60–99	0–50	NA	NA
Clarifier	30–60	20–50	NA	50–80	60–95	0–50	NA	NA
Dissolved air flotation	20–70	10–60	NA	50–85	70–85	10–75	NA	NA
Oxidation pond	40–95	30–65	60	20–70	50–90	60–99	0–15	70–100
Aerated lagoon	75–95	60–85	NA	40–65	70–90	90–99	10–45	95–100
Activated sludge	80–99	50–95	40–90	60–85	80–99	95–99+	33–99	97–100
Trickling filter	60–85	30–70	NA	60–85	50–80	70–98	15–90	70–100
Granular media filter	NA	NA	50–65	75–95	65–95	5–20	NA	NA
Granular activated carbon	91–98	86–94	50–80	60–90	70–95	90–99	33–87	NA
Powdered activated carbon	80–99	50–98	NA	NA	NA	NA	NA	NA

NA = Data not available.

1.4.5 Residuals management

Radian Corporation (1979) reviews the solid wastes that are generated by many operations within a refinery. These residuals can be a significant portion of the pollutant load generated by this industry. Solid wastes include oil skimmings, wastes from tank bottom cleaning, filter clays and water and wastewater treatment sludges. Table 1.12 presents an estimate of the total annual solid

Table 1.12 Solid waste generation (tonnes per year) of the entire USA petroleum refining industry (Radian, 1979)

Stream	Total generation	95% confidence interval for total generation
Slop oil emulsion solids	59 380	9 690– 28 720
Silt from stormwater runoff	139 900	16 390– 923 200
Exchanger bundle cleaning sludge	690	90– 3 700
API separator sludge	51 090	21 500– 94 090
Non-leaded gasoline tank bottoms	900	80– 72 200
Crude tank bottoms	14 450	2 470– 56 460
Other storage tank bottoms	21 500	230– 149 800
Leaded gasoline tank bottoms	4 400	1 070– 13 920
Dissolved air flotation skimmings	76 110	24 670– 174 400
Kerosene filter clays	2 470	550– 8 100
Other filter clays	44 930	4 230– 378 800
HF alkylation sludge	34 180	2 100– 421 100
Waste bio-sludge	97 790	22 380– 361 200
Once-through cooling-water sludge	2 470	0– 47 570
FCC catalyst	99 370	37 000– 123 300
Coke fines	35 770	880– 815 800
Spent amines	18	0– 350
Salts from regeneration	—	—
Ship and barge ballast	—	—
Other	2 290	140– 23 300
Total	687 708	143 470–3 695 950

waste generation by this industry in the USA. Twenty residual streams are included in this table.

Solid waste streams can be handled by way of three approaches either separately or in combination, as follows:

- Source control
- Treatment
- Disposal

Source control

Source control can be defined as the reduction or containment of plant emissions at their point of generation. Instead of continually investing in more extensive pollution control equipment to meet new regulations, refiners are making significant moves to reduce wastes generated. By continuously implementing new source control methods, refiners anticipate the constantly changing regulations and minimize the effort and cost of compliance. One approach to source control is the addition or alteration of equipment. For instance, installation of tank mixers in crude and other storage tanks minimizes the deposition of bottoms sludge.

Another approach is implementation of operational or procedural changes. Two examples are:

- Shutdown planning — the amount and quality of oily wastes is predetermined and handled appropriately in the event of an equipment shutdown.
- Internal refinery permit system — an oily stream can be released to the oily drains only with the approval of the plant environmental department. An internal charging system can promote interest in reducing waste discharges.

With the increased promulgation of environmental regulations and rising cost of pollution control, in many cases the most cost-effective approach to management of refinery sludges is to locate and reduce emissions at the source.

Treatment

Until the last few years, most refiners considered solid waste disposal as a low priority item and a profit drain. Consequently, convenience rather than environmental concern dictated use of sludge pits and lagoons, usually located at the back of refinery property. Little attention was given to leachate, air emissions or ultimate disposal of pollutants contained in these pits. If solid wastes were consigned to a contractor, the material was considered properly disposed of as soon as it left the refinery.

The abundance of environmental regulations has forced refiners to evaluate critically their pollution abatement systems. One important aspect of these systems is the management of solid waste. The following are technologies available to this industry.

(1) Proven treatment technologies currently in use in refineries:
- Vacuum filtration
- Pressure filtration
- Centrifugation
- Dewatering lagoons
- Aerobic digestion
- Anaerobic digestion
- Chemical fixation

(2) Proven treatment technologies potentially usable in refineries:
- Sand drying beds
- Wet-air oxidation
- Composting

Final disposal

Much indecision on final disposal techniques currently exists in the refining industry in the USA. Most of this is due to pending EPA regulations, primarily the Resource Conservation and Recovery Act (RCRA) regulations for environmentally acceptable disposal methods, combined with the potential classification of many refinery wastes as hazardous. The disposal practices used by this industry at present are as follows:

- Incineration
- Landfarming
- Landfill
- Ponding
- Disposal wells
- Ocean disposal

Only landfarming can truly be considered an *ultimate* disposal technique because the wastes are degraded into matter that becomes a part of the soil. Although anaerobic degradation theoretically occurs in landfills, some of those that have been recently reopened showed no significant deterioration of the oily wastes. This is to be expected, since the microbial degradation pathway for hydrocarbons is known to involve the insertion of oxygen into the chemical structure. Hence, anaerobic decomposition of hydrocarbons is essentially nil. Ponding, disposal wells and ocean disposal are merely storage techniques with no significant waste degradation. Incineration provides for destruction of a high percentage of the organics, but this process generates air pollutants and ash residues that need disposal.

Ocean disposal, deep well injection and ponding are being carefully reconsidered for potential adverse environmental impact. The use of landfarming, however, is increasing significantly where land is available. The possibility of farming organic wastes from retired refinery sludge pits after recovering and recycling oil is being seriously considered.

Contracting wastes to local municipalities or private disposal firms is prevalent in the petroleum refining industry. Various proposed regulations will restrict this practice, which may increase on-site disposal activities.

1.5 Costs

The cost of the purchase and installation of treatment facilities is site-specific and is affected by plant layout, land availability and local weather conditions. Therefore, only generalized costs can be estimated for the industry as a whole. The costs presented below do not include the cost of land or take account of any site-specific problem that could affect the cost. Costs have been taken from EPA's development document for petroleum refining (EPA, 1979).

In addition, depreciation and the cost of money vary between refineries because of different accounting procedures and interest rates. Therefore, these factors have not been included as operating costs.

1.5.1 Recycle and reuse

Recycle and reuse of process wastewaters can involve many techniques. As discussed in the previous section, these techniques include the following:

- Collection and reuse of steam condensate
- Collection and reuse of additional cooling water
- Reuse of sour water in the desalter
- Reuse of treated effluent

All of these techniques primarily require the use of pumps and piping. In some cases, oil removal may be required, or water softening may be appropriate. Table 1.13 presents capital and operating costs for pumps and piping, per 1.6 km of piping required. Six flow rates between $2.3 \, m^3 \, h^{-1}$ and $800 \, m^3 \, h^{-1}$ are presented.

Table 1.14 presents capital and operating costs for water-softening equipment. Costs include chemical addition, clarification and filtration. As can be seen, chemical costs account for a major portion of the annual operating expenditure.

1.5.2 In-plant control

Sour water stripping

Figure 1.12 presents capital costs for sour water strippers. The three cost curves presented relate to units designed for high nitrogen crudes, standard hydrogen sulphide strippers and modification costs for ammonia removal. Table 1.15 presents typical stripper operating costs for three sizes of refinery. Energy consumption, as reflected in the cost of steam, accounts for the high operating costs of this type of wastewater treatment.

Treatment of cooling tower blowdown

Cooling tower blowdown may contain hexavalent chromium if this metal is used for corrosion control. Treatment of this pollutant generally consists of chemical reduction followed by precipitation and clarification. Table 1.16 presents capital and operating costs for this type of chromium removal system. In general, chromium treatment is provided in batch systems operated below $80 \, m^3 \, d^{-1}$. This type of treatment operation is very labour-intensive.

1.5.3 End-of-pipe treatment

Dissolved air flotation

After treatment in an API separator, which is generally considered a refinery process unit for the recovery of oil, oil and grease can be further removed in a

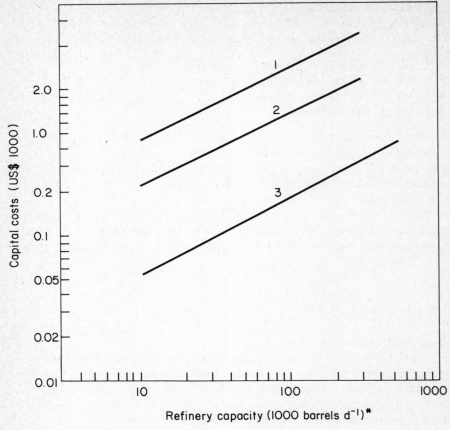

Fig. 1.12 Sour water stripping — capital costs
    1 High nitrogen crude installations
    2 New installations for H_2S stripping
    3 Revisions for NH_3 stripping
* 1 barrel per day = $0.208 \, m^3$ per day

dissolved air flotation (DAF) unit. Table 1.17 presents capital and operating costs for four DAF systems. These costs are based upon steel units, complete with pressurized tank(s), recycle pump(s) and chemical feed equipment.

Biological treatment

The biological treatment system most commonly used in this industry is the activated sludge process. Table 1.18 presents capital and operating costs for activated sludge systems over a wide range of refinery wastewater flows. These costs are derived from a 1974 US EPA study (EPA, 1974a) and are presented in 1972 US dollars.

Table 1.13 Wastewater recycle — capital and operating costs (US$ per 1.6 km (1 mile) — Reference base 1976)

	Flow rate					
	$2.3\ m^3\ h^{-1}$	$16\ m^3\ h^{-1}$	$80\ m^3\ h^{-1}$	$160\ m^3\ h^{-1}$	$320\ m^3\ h^{-1}$	$800\ m^3\ h^{-1}$
CAPITAL COSTS						
Piping:						
Piping, installed, per 1.6 km	$32 000	$53 000	$100 000	$135 000	$175 000	$243 000
Misc. costs (15%)	5 000	8 000	15 000	20 000	26 000	36 000
Total constructed cost per 1.6 km	37 000	61 000	115 000	155 000	201 000	279 000
Engineering (15%)	6 000	9 000	18 000	23 000	30 000	42 000
Contingency	7 000	10 000	17 000	22 000	29 000	42 000
Piping — total capital cost per 1.6 km	$50 000	$80 000	$150 000	$200 000	$260 000	$363 000
Pumps:						
Pumps and associated equipment installed (10% of piping cost)	5 000	8 000	15 000	20 000	26 000	37 000
Total capital cost per 1.6 km	$55 000	$88 000	$165 000	$220 000	$286 000	$400 000
(Minimum pumping costs regardless of distance)	5 000	6 000	12 000	18 000	24 000	40 000
ANNUAL OPERATING COSTS						
Pumping costs per 1.6 km per year	$ 100	$ 700	$ 2 600	$ 4 500	$ 9 200	$ 24 300
Maintenance (1.5% of capital costs) per 1.6 km per year	800	1 300	2 500	3 300	4 300	6 000
Total annual operating cost	$ 900	$ 2 000	$ 5 100	$ 7 800	$ 13 500	$ 30 300

Table 1.14 Water softening of recycled wastewater — capital and operating costs (US$ — Reference base 1976)

	Flow rate					
	$2.3 \text{ m}^3 \text{ h}^{-1}$	$16 \text{ m}^3 \text{ h}^{-1}$	$80 \text{ m}^3 \text{ h}^{-1}$	$160 \text{ m}^3 \text{ h}^{-1}$	$320 \text{ m}^3 \text{ h}^{-1}$	$800 \text{ m}^3 \text{ h}^{-1}$
CAPITAL COSTS						
Solids contact clarifier	$25 000	$30 000	$45 000	$65 000	$80 000	$125 000
Chemical feed system(s)	5 000	7 000	10 000	15 000	25 000	50 000
Filter unit	15 000	25 000	30 000	40 000	80 000	150 000
Subtotal	45 000	62 000	85 000	120 000	185 000	325 000
Auxiliary equipment	5 000	8 000	10 000	15 000	20 000	35 000
Total capital cost	50 000	70 000	95 000	135 000	205 000	360 000
Installation (50%)	25 000	35 000	50 000	70 000	100 000	180 000
Total constructed cost	75 000	105 000	145 000	205 000	305 000	540 000
Engineering	15 000	20 000	25 000	30 000	45 000	80 000
Contingency	15 000	20 000	25 000	30 000	45 000	80 000
Total capital cost	$105 000	$145 000	$195 000	$265 000	$395 000	$700 000
ANNUAL OPERATING COSTS						
Lime and caustic feed (based on 500 mg l^{-1} hardness)	$1 000	$7 000	$35 000	$70 000	$140 000	$350 000
Misc. power	100	700	3 500	7 000	14 000	35 000
Labour ($10 per man-hour)	4 000	5 000	6 000	7 000	8 000	9 000
Maintenance (3% of total capital cost)	3 200	4 400	5 900	8 000	12 000	21 000
Total annual cost	$8 300	$17 100	$50 400	$92 000	$174 000	$415 000

Table 1.15 Sour water strippers — annual operating costs (US$ — Reference base 1976)

	Flow rate		
	4200 m^3 d^{-1} (20 000 bbl d^{-1})	20 000 m^3 d^{-1} (95 000 bbl d^{-1})	31 000 m^3 d^{-1} (150 000 bbl d^{-1})
Steam: $3.31 per 1000 kg ($1.50 per 1000 lb)	$25 000	$310 000	$430 000
Pumping: 709 kW m^{-3} s^{-1} $0.04 per kWh	500	5 000	8 000
Labour ($\frac{1}{2}$ man-year)	10 000	10 000	10 000
Maintenance (3% of total capital cost)	13 000	28 000	35 000
Total annual cost	$48 500	$353 000	$483 000

Filtration

Filtration is the most cost-effective polishing system for improving effluents from biological treatment systems. Table 1.19 presents capital and operating costs for dual media filtration systems of various sizes. The costs are based on steel shell gravity filters that contain integral backwash storage compartments. Costs include pumps, immediate piping, gravity splitter box(es) and installation.

Activated carbon

Table 1.20 presents costs for powdered activated carbon treatment systems and Table 1.21 presents capital and operating costs for granular activated carbon treatment systems. Costs for powdered carbon systems are based upon a carbon dosage of 80 mg l^{-1} and include interconnecting piping. The granular carbon systems include on-site regeneration equipment for systems over 3800 m^3 d^{-1}. In both types of carbon treatment system, carbon makeup represents the most significant operating cost.

1.6 Summary

This chapter has provided information on the processes utilized and products produced by the petroleum refining industry, and the characteristics of the wastewaters emanating from these processes. It has also addressed the discharge standards relating to petroleum refineries generally throughout the world, the control and treatment technologies utilized for water pollution abatement in the industry, and the costs associated with these technologies.

The petroleum refining industry is extremely complex and has been undergoing significant change in the recent past in several respects. Since the 1970s, concern about and awareness of hazards to the environment throughout the developed countries of the world has caused the industry to institute extensive control and treatment practices for its wastewaters. A great deal of

Table 1.16 Chromium removal systems — capital and operating costs (US$ — Reference base 1976)

	Flow rate					
	3.8 m³ d⁻¹	38 m³ d⁻¹	380 m³ d⁻¹	3800 m³ d⁻¹	19 000 m³ d⁻¹
CAPITAL COSTS					
Detention tank	$ 100	$ 1 000	$ 5 000	$ 20 000	$ 50 000
Chemical feed systems	5 000	15 000	30 000	40 000	45 000
Automatic controls	—	10 000	10 000	10 000	10 000
Solids contact clarifier pumps	25 000	25 000	35 000	80 000	155 000
Total equipment cost	30 100	51 000	80 000	150 000	260 000
Installation (50%)	15 000	25 500	40 000	75 000	130 000
Total constructed cost	45 100	76 500	120 000	225 000	390 000
Engineering	6 950	11 750	17 500	37 500	60 000
Contingency	6 950	11 750	17 500	37 500	60 000
Total capital cost	$59 000	$100 000	$155 000	$300 000	$510 000
ANNUAL OPERATING COSTS					
Sulphur dioxide	$ 16	$ 160	$ 1 600	$ 16 000	$ 80 000
Acid	4	40	400	4 000	20 000
Caustic	130	1 300	13 000	130 000	620 000
Mixing	70	70	400	4 000	21 000
Pumping	Negligible	10	100	1 000	5 000
Labour	5 200	5 200	5 200	10 000	20 000
Maintenance (3% of total capital cost)	1 780	3 000	4 800	9 000	16 000
Total annual cost	$ 7 200	$ 9 780	$ 25 500	$174 000	$782 000

Table 1.17 Dissolved air flotation — capital and operating costs (US$ — Reference base 1976)

	Flow rate			
	300 m^3 d^{-1}	3800 m^3 d^{-1}	16 700 m^3 d^{-1}	23 600 m^3 d^{-1}
CAPITAL COSTS				
Dissolved air flotation unit with instruments and controls	$35 000	$ 80 000	$130 000	$150 000
Chemical injection equipment	15 000	30 000	45 000	55 000
Subtotal	50 000	110 000	175 000	205 000
Piping (10%)	5 000	11 000	17 500	20 500
Total equipment cost	55 000	121 000	192 500	225 500
Installation (50%)	27 500	60 500	96 500	112 500
Total constructed cost	82 500	181 500	289 000	338 000
Engineering (15%)	12 500	27 300	43 500	51 000
Contingency	15 000	26 200	42 500	51 000
Total capital cost	$110 000	$235 000	$375 000	$440 000
ANNUAL OPERATING COSTS				
Chemicals				
Alum	$ 1 000	$ 14 000	$ 62 000	$ 86 000
Polyelectrolyte	500	6 000	27 000	39 000
Power (electricity)				
DAF unit requirements	1 400	8 000	35 000	50 000
Chemical feed pumps and mixers	200	400	2 000	3 000
Labour (0.25 man-years)	5 000	5 000	5 000	5 000
Depreciation (20%)	22 000	47 000	75 000	88 000
Maintenance (3% of total capital cost)	3 500	7 000	11 000	13 000
Total annual cost	$ 33 600	$ 87 400	$217 000	$284 000

emphasis has been placed on in-plant controls and water conservation and reuse, in that these methodologies are the most cost-effective means of controlling wastewater discharges. It is expected that this trend will continue, with an ever-increasing emphasis on conservation, wastewater recycling and reuse, and in-plant controls to minimize the quantity and pollutant characteristics of the wastewater generated.

In terms of residuals management, it is expected that increased awareness and concern will be focused on the ultimate disposal of both hazardous and non-hazardous residuals throughout the coming decade. In the petroleum refining industry, it is anticipated that the trend toward land treatment, specifically landfarming, will increase and receive even more attention in the future. As the industry advances in processing technology, the management of its wastewaters and residuals is expected to move forward in concert.

Table 1.18 Activated sludge systems — capital and operating costs (US$ — Reference base 1972)

	Flow rate					
	510 m^3 d^{-1}	1100 m^3 d^{-1}	4100 m^3 d^{-1}	22 000 m^3 d^{-1}	41 000 m^3 d^{-1}	57 000 m^3 d^{-1}
Capital costs	$290 000	$400 000	$1 120 000	$4 070 000	$7 720 000	$10 100 000
Operating costs	31 000	45 000	119 000	484 000	840 000	1 239 000

Table 1.19 Tertiary dual media filtration — capital and operating costs (US$ — Reference base 1976)

	Flow rate					
	380 m^3 d^{-1}	3800 m^3 d^{-1}	19 000 m^3 d^{-1}	38 000 m^3 d^{-1}	76 000 m^3 d^{-1}
CAPITAL COSTS					
Filtration units installed	$25 000	$100 000	$250 000	$350 000	$600 000
Interconnecting piping, installed	3 000	10 000	25 000	35 000	60 000
Pumps, installed	5 000	15 000	42 000	60 000	100 000
Total installed cost	33 000	125 000	317 000	451 000	770 000
Engineering	6 000	20 000	49 000	69 500	115 000
Contingency	6 000	20 000	49 000	69 500	115 000
Total capital cost	$45 000	$165 000	$415 000	$590 000	$1 000 000
ANNUAL OPERATING COSTS					
Pumping	$ 140	$ 1 400	$ 7 000	$ 14 000	$ 28 000
Labour	4 000	5 000	6 000	7 000	8 000
Maintenance (3% of capital cost)	1 400	5 000	12 500	18 000	30 000
Total annual cost	$ 5 540	$ 11 400	$ 25 500	$ 39 000	$ 66 000

Table 1.20 Powdered activated carbon — dosage 80 mg l^{-1} — capital and operating costs (US$ — Reference base 1976)

	Flow rate				
	380 m^3 d^{-1}	3800 m^3 d^{-1}	19 000 m^3 d^{-1}	38 000 m^3 d^{-1}	76 000 m^3 d^{-1}
CAPITAL COSTS					
Powdered carbon feed system	$10 000	$30 000	$ 45 000	$ 60 000	$ 100 000
Piping	1 000	3 000	4 500	6 000	10 000
Total equipment cost	11 000	33 000	49 500	66 000	110 000
Installation (50%)	6 000	16 500	24 800	33 000	55 000
Total constructed cost	17 000	49 500	74 300	99 000	165 000
Engineering	9 000	10 000	11 350	15 500	25 000
Contingency	9 000	10 000	11 350	15 500	25 000
Total capital cost	$35 000	$69 500	$ 97 000	$130 000	$ 215 000
ANNUAL OPERATING COSTS					
Carbon make-up	$ 7 400	$74 000	$370 000	$740 000	$1 480 000
Miscellaneous power requirements	1 000	2 000	5 000	8 000	15 000
Labour ($10 per man-hour)	4 000	5 400	9 400	12 400	19 400
Maintenance (3% of total capital cost)	1 000	2 000	3 000	4 000	6 600
Total annual cost	$13 400	$83 400	$387 400	$764 400	$1 521 000

Table 1.21 Granular activated carbon — capital and operating costs (US$ — Reference base 1976)

	Flow rate				
	380 m³ d⁻¹	3800 m³ d⁻¹	19000 m³ d⁻¹	38000 m³ d⁻¹	76000 m³ d⁻¹
CAPITAL COSTS					
Activated carbon units	$ 50000	$ 325000	$1500000	$2600000	$ 5000000
Pumping and misc. equip. (10%)	5000	32500	150000	260000	500000
Piping (10%)	5000	32500	150000	260000	500000
Total equipment cost	60000	390000	1800000	3120000	6000000
Installation (50%)	30000	195000	900000	1560000	3000000
Total constructed cost	90000	585000	2700000	4680000	9000000
Engineering	40000	85000	400000	710000	1350000
Contingency	20000	80000	400000	710000	1350000
Subtotal	150000	750000	3500000	6100000	11700000
Activated carbon regeneration system (installed)	—	300000	450000	600000	750000
Contingency (for utility hook-up, etc.)	—	60000	100000	120000	150000
Engineering for carbon regeneration system	—	50000	50000	80000	100000
Total capital cost	$150000	$1160000	$4100000	$6920000	$12700000
ANNUAL OPERATING COSTS					
Carbon makeup	$ 28000	$ 28000	$ 137000	$ 275000	$ 550000
Furnace power	—	19000	27000	46000	82000
Pumping	500	5000	25000	50000	100000
Labour ($10 per man-hour)	21000	98000	105000	115000	125000
Maintenance (3% of total capital cost)	4500	35000	123000	208000	381000
Total annual cost	$ 54000	$ 185000	$ 417000	$ 694000	$ 1238000

References

Anon. (1977) *Gas Processing Refining and Worldwide Directory, 1976–77*, 34th edn, Petroleum Publishing Company, Tulsa, Oklahoma.
Anon. (1978) Why Third World nations ignore pollution, *Business Week*, 10 July, 90D.
Anon. (1979) *Quality of the Environment in Japan — 1979*, Environment Agency, Japan.
Anon. (1980) Refining report (March 24, 1980), *Oil Gas J.*, **78**(12), 75–159.
American Water Works Association (1971) *Water Quality and Treatment*, McGraw–Hill, New York.
Beychok, M.R. (1967) *Aqueous Wastes from Petroleum and Petrochemical Plants*, Wiley, London.
Bland, W. and Davidson, R. (1967) *Petroleum Processing Handbook*, McGraw–Hill, New York.
EPA (1974a) *Development Document for Effluent Limitations Guidelines and New Source Performance Standards for the Petroleum Refining Point Source Category*, EPA 440/1–74–014–a, US Environmental Protection Agency, Washington, DC.
EPA (1974b) *Development Document for Effluent Limitations Guidelines and New Source Performance Standards for the Copper, Nickel, Chromium, and Zinc Segment of the Electroplating Point Source Category*, EPA 440/1–74–003–a, US Environmental Protection Agency, Washington, DC.
EPA (1976) *Draft Supplement for Pretreatment to the Development Document for the Petroleum Refining Industry, Existing Point Source Category*, EPA 400/1–76/083, US Environmental Protection Agency, Washington, DC.
EPA (1977) *Federal Guidelines: State and Local Pretreatment Programs*, EPA 430/9–76–017a, US Environmental Protection Agency, Washington, DC.
EPA (1978) *Innovative and Alternate Technology Assessment Manual*, EPA 430/9–78–009, US Environmental Protection Agency, Washington, DC.
EPA (1979) *Development Document for Proposed Effluent Limitations Guidelines, New Source Performance Standards and Pretreatment Standards for the Petroleum Refining Point Source Category*, EPA 440/1–79/014–b, US Environmental Protection Agency, Washington, DC.
EPA (1980) *Treatability Manual, Technologies for Control/Removal of Pollutants*, Vol. III, EPA 600/8–80–042c, US Environmental Protection Agency, Washington, DC.
Fern, G.R.H. (1974) *Petroleum Refinery Water Usage*, PACE–EPS Technology Transfer Seminar, Environment Canada, Edmonton.
Finelt, S. and Crump, J.R. (1977) Pick the right water reuse system, *Hydrocarbon Process.*, **56**(10), 111–120.
Grieves, C.G., Stenstrom, M.K., Walk, J.D. and Grutsch, J.F. (1977) *Effluent Quality Improvement by Powdered Activated Carbon in Refinery Activated Sludge Processes*, presented at the 42nd Midyear Meeting, API Refining Department, 11 May, Chicago, Illinois.
Guthrie, V.B. (1960) *Petroleum Products Handbook*, McGraw–Hill, New York.
Meyerson, A. (1980) Japan: environmentalism with growth, *Wall Street Journal*, 5 September.

Nelson, W.L. (1969) *Petroleum Refinery Engineering*, McGraw–Hill, New York.
OECD (1979) *The State of the Environment in OECD Member Countries*, Organization for Economic Cooperation and Development, Paris.
Orr, O. (1974) *Refinery Processes, Petroleum Refinery Technology and Pollution Control*, PACE–EPS Technology Transfer Seminar, Environment Canada, Edmonton.
Radian Corporation (1979) *Refinery Waste Disposal Screening Study*, NTIS Rep. PB 13–299–351, US Environmental Protection Agency, Washington, DC.
Reed, D.T., Klen, E.F. and Johnson, D.A. (1977) Use of side stream softening to reduce pollution, *Hydrocarbon Process.*, **56**(11), 339–342.
Rizzo, J.A. (1976) *Case History: Use of Powdered Activated Carbon in an Activated Sludge System*, presented at the First Open Forum on Petroleum Refinery Wastewaters, January, Tulsa, Oklahoma.
Thibault, G.T., Tracy, K.D. and Wilkinson, J.B. (1977) *Evaluation of Powdered Activated Carbon Treatment for Improving Activated Sludge Performance*, presented at the 42nd Midyear Meeting, API Refining Department, 11 May, Chicago, Illinois.
Vesik, H. (1974) *Refinery Operations, Petroleum Refinery Technology and Pollution Control*, PACE–EPS Technology Transfer Seminar, Environment Canada, Edmonton.

2 The treatment of wastes from the synthetic fuels industry

R D Neufeld, *Department of Civil Engineering, University of Pittsburgh, Pennsylvania*

2.1 Introduction

The history of man's civilization is tied inextricably to the availability of inexpensive and plentiful sources of energy. During the early ages of man, wood was a common source of energy. With the advent of the Industrial Revolution, wood and coal were utilized to feed the furnaces of industrial progress. During the first half of the twentieth century coal was used as the principal source of power generation for industry and cities alike. The 1940s saw a boom in the production of petroleum, with oils from the Middle East slowly replacing North American coal. Nuclear power is today considered a plentiful source of cheap and available energy for continued progress through the twenty-first century.

The difficulty lies in the short term – the period covering the phasing-in of comprehensive nuclear facilities. Because of the political and social conditions of the 1980s, it appears that oil-based economies will find extreme difficulty in maintaining an adequate and secure energy supply. It is clear that the USA and many other Western countries must turn back to their abundant coal and oil shale reserves, which have been lying dormant for many years, as the fuel alternative for the short term (30–70 years).

The goal of the synthetic fuels industry is simply to convert the solid energy resource, coal, into more readily usable forms such as liquid or gaseous fuels. Additional possibilities exist in the conversion of synfuels gases into petroleum-like products such as waxes, oils and the whole array of products that are now derived from petroleum.

A viable synthetic fuels industry is a distinct possibility: such an industry provided fuel for the Axis powers during World War II. A synthetic fuels industry based upon similar German technology is now providing energy for South Africa, which currently finds difficulty in importing sufficient petroleum. The USA and other Western countries are currently devoting large sums of money and technical resources to further development and refining of second, third and further generations of coal conversion technology. The basic differences between these advanced technologies and first-generation technology lie in process efficiency, materials handling and the key area of environmental engineering.

Environmental engineering considerations
Because the synthetic fuels industry is relatively new, it is more profoundly affected by public concern for the environment than are the more established industries.

Historically, 'sanitarians' were concerned with biological vectors of disease. Typical objects of concern were diseases such as cholera, typhoid and other biologically-based waterborne diseases which we now see have relatively direct causes. Chlorination and good sanitary practices have helped to reduce the incidence of many waterborne diseases in the Western world.

A major concern for environmental engineers during the late 1960s and early 1970s was the preservation of the quality of watercourses for their aesthetic and recreational value. The late 1970s and 1980s have seen a rapid growth in the availability of advanced analytical tools for the measurements of trace organics and heavy metals. The availability of trace substance data has grown along with the generation of information on cause and effect of chemical vectors of disease. Relatively recent questions of carcinogenicity, mutagenicity and teratogenicity have raised doubts and fears in the minds of the public which may be expected to have a heavy impact on the development of any organic-based industry; the synthetic fuels industry is no exception. Thus, while the initial industrial development of synthetic fuels was prompted by considerations of 'national emergency', today's broad-scale synthetic fuels industry must come to grips with the public's real fears and perceived questions of environmental quality and environmental health.

The purpose of this chapter is to provide a rational basis for decision-making with regard to environmental engineering in the synthetic fuels industry.

2.2 Coal conversion processes

2.2.1 General considerations

Coal conversion systems may be readily classified into two generic categories: coal gasification and coal liquefaction. Coal liquefaction processes may be subdivided into two further categories: direct liquefaction and indirect liquefaction. All direct liquefaction techniques include basic hydrogenation reactions, with or without catalytic aids, whereby coal is dissolved in a hydrogen-rich solvent; this results in an increased H/C ratio and a yield of offgases, liquid fuel and solid waste residuals consisting of heavy organics, inorganics and sulphur products.

Indirect liquefaction consists of coal gasification followed by gas cleaning, the last step in the overall process being catalytic reaction of synthetic gases to form end-products other than methane. Gasification, in fact, is an integral part of direct liquefaction processes as well. In this case the solid waste residual from the first step direct liquefaction, which is about 50% carbon by weight, is further gasified to produce hydrogen via steam decomposition as is necessary for use in the preceding direct liquefaction step.

The three ingredients needed to produce gas, or liquids, from coal are carbon, hydrogen and oxygen. Oxygen is required to support partial combustion to provide energy for the endothermic gasification reactions. Coal provides the carbon source, while steam provides the hydrogen source. Oxygen is provided either by air or via an oxygen plant within the gasification complex. If oxygen is provided from an air supply, the resulting product is known as a low BTU gas and contains nitrogen as a diluent in the final end-product. If an oxygen plant is substituted for the oxygen source, the final product will be considered a medium BTU gas, i.e., without a nitrogen diluent. If a high BTU gas is required, formation of CH_4 is required and nitrogen as a diluent is not permitted. In such cases, medium BTU gas is further upgraded via a 'shift' reaction to upgrade the hydrogen/carbon ratio for the catalytic production of methane.

2.2.2 The integrated coal refinery

An integrated coal conversion facility (coal refinery) comprises the following operations.

Coal storage. An inventory of about 90 to 180 days worth of coal is normally provided at the site of the coal conversion facility. Runoff from the coal storage area represents one source of wastewater that must be accommodated within the plant complex.

Coal preparation. Run-of-mine coal must be suitably crushed and screened to produce a size-distribution compatible with the gasification operations. In addition, boney (clay-like) materials and iron pyrites must be separated from the coal before it is fed to the gasifier. The technologies and wastewaters in the coal preparation area are quite similar to those currently encountered in preparation of coal for coal combustion–electric utility operations and need not be discussed further.

Coal gasification. The operations that give rise to the characteristic wastewaters and residuals from the coal conversion process are described below.

2.2.3 Coal gasification

Gas purification

The offgas from a gasification operation consists of carbon monoxide (CO), hydrogen (H_2), ammonia, sulphide and other trace organics. The purpose of the gas purification operation is to take scrubbed gases and remove residual sulphur to give a usable gas product. In a sense, the gas purification operation may be looked upon as a desulphurization step. This step makes gasification attractive for cheaper high sulphur coals in electricity generation via combined cycle systems as well as via direct methene generation.

Gas upgrading

Gas upgrading operations include steps, such as 'shift conversion', which convert the hydrogen and carbon monoxide synthesis gas into a molar ratio suitable for subsequent methanation followed by drying and compression to pipeline standards. Depending upon the wastewater treatment scheme, treated effluents from the shift conversion area may or may not be contaminated with organics. Effluents from the methanation and drying steps are usually relatively pure and may be used as boiler feed-water within the plant complex. Additional details of these streams are given below.

The reactions inside the gasification reactor consist of the following steps.

Drying. Coal is dried within the gasification step by contact with hot gases in the reactor. Coals with high moisture content, such as lignite coals, require more energy for the drying step than do other types of coals. Moisture contained within the coal ends up in the wastewater treatment plant.

Devolatilization. Volatile organic fragments are 'evaporated' into the gas phase stream. In fixed-bed systems, such devolatilization may take place above the coal bed in the free-fall zone of the system.

The gasification reaction. The reactions involved are:

$$\text{Coal} + H_2 \rightarrow CH_4 + C \quad \text{devolatilization-pyrolysis} \quad (1)$$

$$C + 2H_2 \rightarrow CH_4 + \text{Heat} \quad \text{gasification} \quad (2)$$

$$C + H_2O \rightarrow CO + H_2 \quad \text{gasification} \quad (3)$$

$$C + O_2 \rightarrow CO_2 + \text{Heat} \quad \text{combustion} \quad (4)$$

$$C + \tfrac{1}{2}O_2 \rightarrow CO + \text{Heat} \quad \text{combustion} \quad (5)$$

Reaction (3) is highly endothermic and will not occur unless sufficient heat is present. Accordingly, reaction (4) is incorporated, which provides a partial combustion of coal, producing carbon dioxide and sufficient energy to drive the above reaction. As outlined above, if air is used as the oxygen carrier, the product gas will have a relatively low heating value 5600–11 000 kJ m^{-3} and is designated a low BTU gas. The use of pure oxygen results in a medium BTU gas (11 000–15 000 kJ m^{-3}; 300–400 Btu/SCF) with no nitrogen as diluent. In both cases carbon dioxide and hydrogen sulphide are present in the product gas.

Since the hydrogen necessary for reactions (1) and (2) comes from the decomposition of steam, gasification operations require steam injection as an integral part of the gasification chemistry. Often, excess steam injection is required, since only a portion of the steam is inherently decomposed. This excess goes into the effluent wastewater stream. For example, the Synthane process achieves only 15% steam decomposition (Johnson *et al.*, 1977) the remainder of the steam resulting in a process condensate. Where a high BTU gas of pipeline quality is desired, steps must be taken to methanate the final product. Such steps involve hydrogen-enrichment of the synthesis gas, which is accomplished via the water gas shift reaction:

$$CO + H_2O \rightarrow H_2 + CO_2 + \text{Heat} \tag{6}$$

Reaction (6) is conducted so that the final hydrogen-to-carbon monoxide ratio in the existing stream is at least $3:1$ (the required ratio for conducting further methanation steps). Methanation, if used, is as follows:

$$3H_2 + CO \rightarrow CH_4 + H_2O + \text{Heat} \tag{7}$$

and

$$4H_2 + CO_2 \rightarrow CH_4 + 2H_2O + \text{Heat} \tag{8}$$

The removal of H_2S and other trace sulphur compounds from shifted gases is necessary to prevent poisoning of downstream catalysts in the production of high BTU gas. A nickel-based catalyst which is readily poisoned by sulphur is often used. For indirect liquefaction purposes, sulphur removal is similarly required. The removal of CO_2 is usually necessary if synthetic natural gas of $37\,000\,kJ\,m^{-3}$ (1000 Btu/SCF) is desired. Techniques for the bulk removal of hydrogen sulphide and carbon dioxide are known as acid gas removal processes and often generate additional residuals which must be accommodated in the overall wastewater treatment facility.

Gasifier wastewater condensates

The major aqueous effluent stream to be considered is that derived from the gasification process itself. In an idealized case, sufficient energy is provided via combustion to raise the temperature of coal in the gasifier reactor so that all carbon goes to carbon monoxide, all coal-nitrogen goes to ammonia and all coal-sulphur goes to hydrogen sulphide. Indeed, in gasification facilities utilizing a high value of oxygen/coal, gasification reaction temperatures are extremely high and effluent streams have been noted to be relatively void of organic fractions. Such systems tend, however, to have larger fractions of carbon dioxide in the product gas. It appears that as the oxygen/coal ratio decreases, gasification bed temperatures decrease and the liquid effluents tend to be 'dirtier'.

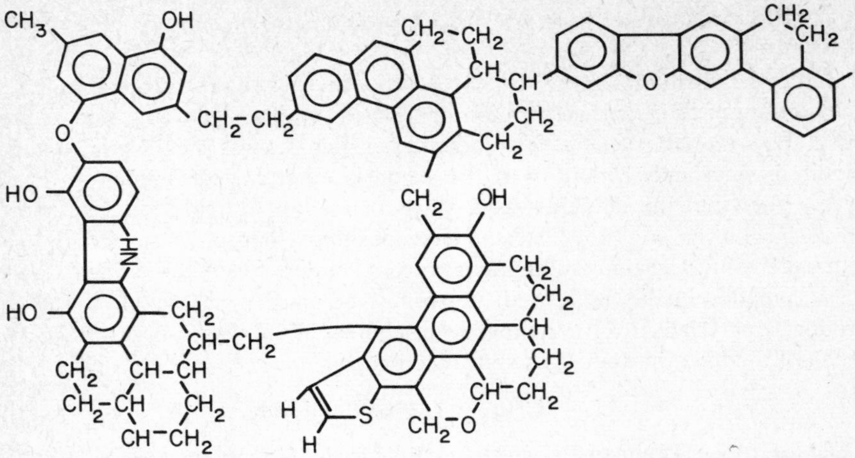

Fig. 2.1 Proposed structure of the model coal molecule (Heredy and Wender, 1980)

An understanding of the nature of coal conversion effluents can be gained from examination of models of coal structures. In summarizing and deriving new models of coal structures, Heredy and Wender (1980) show that bituminous coals contain relatively small condensed aromatic ring systems which consist on average of two to four condensed rings; fluorene- and phenanthrene-type condensed aromatic rings predominate, while the non-aromatic part of the molecule is found to consist mainly of hydroaromatic rings with few alkyl (mainly methyl) groups. Figure 2.1, taken from this study, is a proposed structure of a generalized bituminous coal molecule.

Cleavage of aromatic rings may be expected to result in a phenolic-rich wastewater with the phenolic fraction consisting of simple monohydric phenols, polyhydric phenols and multimethylated phenols. Heredy and Wender (1980) calculate that approximately 70% of coal carbon is aromatic in nature. Figure 2.2 indicates some of the aromatic constituents of the model coal compound. Examination of the proposed coal molecule shows hydroaromatic rings of the type shown in Fig. 2.3 as being predominant. If partial pyrolysis of coal takes place in lower temperature regions in gasification reactors, the result is a partial cleavage of the model coal molecule. We may expect wastewaters from such processes to contain a large array of heterocyclic nitrogen and sulphur-containing substances along with aromatics and relatively shorter-chain fatty acid types of material.

2.3 Coal conversion plant — sources of wastewater

2.3.1 The overall operation

Figure 2.4 depicts a generalized coal conversion process showing sources of effluent streams that may be generated. This generalized process includes a

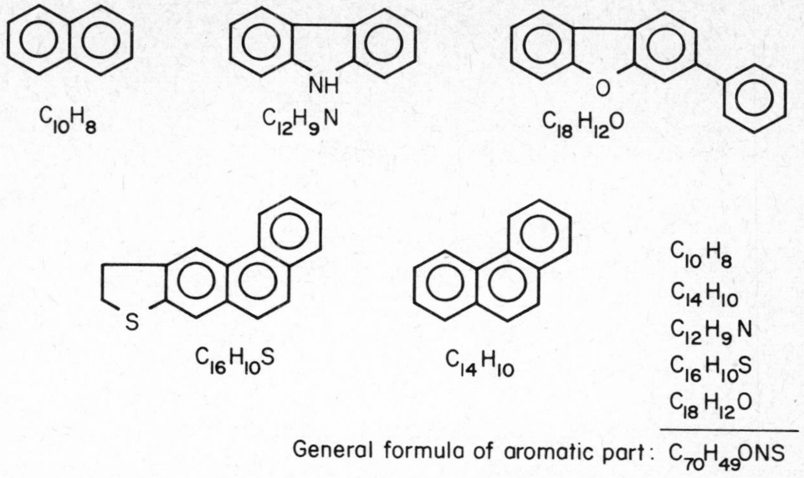

$C_{10}H_8$

$C_{12}H_9N$

$C_{18}H_{12}O$

$C_{16}H_{10}S$

$C_{14}H_{10}$

$C_{10}H_8$
$C_{14}H_{10}$
$C_{12}H_9N$
$C_{16}H_{10}S$
$C_{18}H_{12}O$
_____
General formula of aromatic part: $C_{70}H_{49}ONS$

Fig. 2.2 Aromatic constituents of the model molecule (Heredy and Wender, 1980)

coal preparation area, a coal gasification area, a hot gas clean-up area and a synthesis section where the final hot gas product is either converted to methane or converted to liquid products via a variety of indirect catalytic liquefaction steps. Processes for such liquefaction include the Mobil–M and Fischer–Tropsch processes.

The waste streams of any such facility include the following.

Run-off from on-site coal storage, handling and preparation of equipment
Such effluents are most likely to include suspended solids consisting of coal fines and grit, as well as dissolved organics and inorganics leached from the coal itself. The composition of such run-off material is likely to be similar to that of acid mine drainage if high sulphur coals are used. The key factors influencing such characteristics include the coal type and the conditions of wastewater contact with coal, such as residence times and temperatures. Rainfall rates and coal storage and washing practices determine flow rate. Such wastewaters, if acidic, may be used for pH adjustment and dilution within the wastewater treatment train.

Ash quench and sluice waters
These waters result from the water quenching of hot bottom ash residuals from a gasification scheme. This water includes inorganics leached from the ash stream. Such a wastewater stream has been recorded to contain a variety of heavy metal constituents in concentrations sufficiently high to make direct disposal to the environment undesirable.

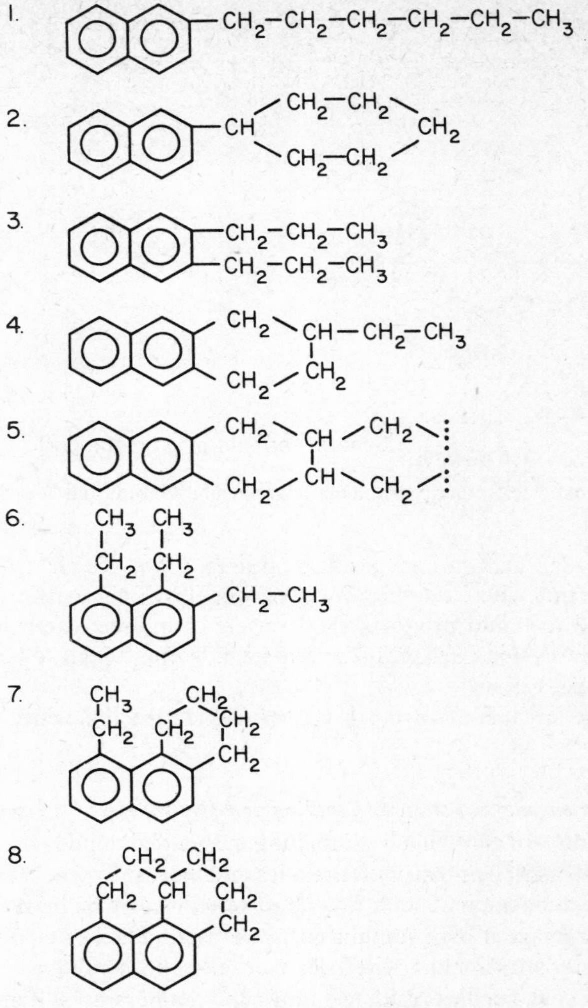

Fig. 2.3 Basic configurations of C_6 aliphatic structures (Heredy and Wender, 1980)

Condensates from raw gas quenching, cleaning and cooling operations
This stream, generically called 'coal conversion wastewater', represents the concentrated phenolic-based wastewater product from coal conversion facilities. This wastewater stream consists of organic and inorganic suspended solids, dissolved organic and inorganic species, ammonia, acid gases, trace metals and a variety of other water-soluble substances. Gasifier operational conditions that influence these stream characteristics include the feed coal properties, gasifier type, oxygen-to-coal ratio, percent steam decomposition in

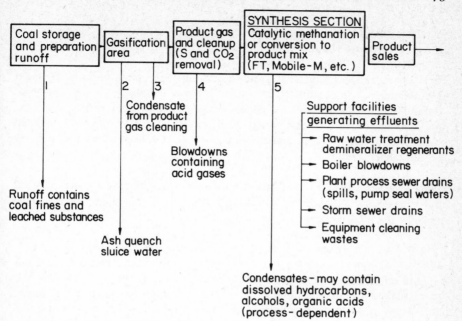

Fig. 2.4 Aqueous wastes generated in coal conversion

the gasification section, and temperatures and pressures within the gas cooling train.

Blowdown solvent and excess wastewaters from the acid gas removal processes
This stream is likely to contain dissolved organics and inorganics including acid gases, solvent and solvent degradation products. As an example, Stretford solutions have been documented in the past to contain high levels of cyanides and thiocyanates as well as other trace organics (Neufeld et al., 1981b). The factors influencing effluent stream characteristics include the nature of the acid gas removal process as well as the properties of the inlet raw gas. Unwanted side reactions related to the inlet gas may necessitate disposal of a blowdown stream.

The synthesis section condensates
This section is likely to include dissolved organics such as alcohols, organic acids and other hydrocarbons. Processes such as Mobil–M and Fischer–Tropsch are expected to produce significant quantities of byproduct condensates similar in composition to the ultimate product. Methanation condensates, on the other hand, contain negligible quantities of dissolved gases and virtually no trace hydrocarbons. These condensates may be considered relatively clean and can go to boiler feed-water preparation operations.

Auxiliary operations

Auxiliary operations necessary for any coal conversion facility are likely to include byproduct streams from the following support areas:

Raw water treatment wastes. These wastes are expected to contain spent demineralizer regenerants which contain considerable levels of dissolved inorganics such as sodium, sulphates, chlorides and other raw water components. The nature of the available natural water stream used for makeup water will determine the characteristics of the dissolved salts in the spent demineralizer stream.

Boiler blowdown. Boiler blowdown includes dissolved inorganics such as silica and trace metals. This is usually a high-quality water stream which can be recycled for feed makeup.

Cooling tower blowdown. Cooling tower blowdowns are expected to include dissolved inorganics and some organic species including makeup water components and treatment chemical residuals. Where wastewater is used as cooling tower makeup, it may be anticipated that evaporation and drift from cooling towers will contain trace nonbiodegradable organics at levels significant to environmental health.

Plant process drains and storm drains. Such drains may contain any of the above material plus cleaning chemicals and metals that may be extracted or leached from equipment surfaces during equipment maintenance. Rainfall rates and plant house cleaning practices affect water flows in storm and plant process drains. Equipment cleaning wastes which may be of negligible flow may nevertheless contain considerable levels of materials potentially toxic or disruptive to subsequent wastewater treatment processes.

As indicated above, the waste streams shown in Fig. 2.4 are typical of any large industrial operation. The waste stream of most concern is that derived from gasifier gas quenching condensates and from indirect liquefaction synthesis operations.

2.3.2 Coal gasification facilities

A quantity of data on the Lurgi coal gasification system has been published. The Lurgi-type system includes a fixed-bed atmospheric or pressurized non-slagging or slagging reactor that uses air or oxygen plus steam as gasification reactants. Partial oxidation pretreatment may be necessary to prevent agglomeration and blocking of the grate. Non-caking or weakly caking high-moisture coal does not need such treatment. A novel water-cooled agitator has been used to gasify mildly caking coals without pretreatment; however, caking bituminous coals require partial oxidation or some other form

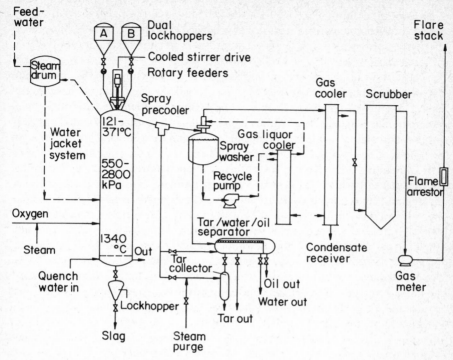

Fig. 2.5 Process flow sheet of GFETC SFBG

of treatment before being processed in the conventional Lurgi-type gasifier. In the Lurgi system the coal is fed into the top of the gasifier and allowed to fall through a rotating grate. Steam and air/oxygen introduced through entry ports below the grate move upwards countercurrently to the coal. With pressurized gasification, a lock hopper system is employed at the top for coal feed and at the bottom for ash removal, thus isolating the pressurized section of the gasifier.

The Grand Forks Energy Technology Center (GFETC) of the US Department of Energy (DOE) is currently (1983) operating the only pilot-plant-sized slagging gasifier in the USA. This facility, which has many features in common with the Lurgi gasification design, has the primary objective of developing basic data required to demonstrate the process within the USA. The Lurgi-type slagging gasifier is being considered by the DOE for funding in a US demonstration-scale plant proposed by American Natural Resources Inc. for Western North Dakota, USA. Figure 2.5 is a process flow diagram of the GFETC slagging fixed-bed gasifier in a complete process train at the North Dakota test site. This facility consists of dual lock hopper arrangements for coal feed in a fixed bed of about 1.8–2.5 m of coal with bottom bed temperature of about 1340°C and top bed temperature in the range of 121°–370°C. The offgas of the gasifier is quenched in a spray washer with recycled wastewater. Since this particular facility uses high-moisture lignite coal as feedstock, the bulk of wastewater produced from this facility originates from the feed coal. Blowdown from the spray washer cycles goes to an oil–water tar separator, from which effluent wastewater is periodically drawn down.

The US DOE also operates a stirred fixed-bed pressurized coal gasification facility of which the design incorporates deviations from that of the conventional Wellman–Galusha gasifier. The Wellman–Galusha unit has been commercially available as an atmospheric coal gasifier since 1941. It is basically a fixed-bed system using steam plus air or oxygen to produce a low or medium BTU gas. The US DOE–Morgantown Energy Technology Center (formerly a US Bureau of Mines Research Center) developed a pressurized version of this facility in 1958, which was designed to reduce steam requirements and increase throughput by reducing coal residence time, thereby increasing unit efficiency and lowering gasification costs. The wastewater condensate train located at the West Virginia, USA, site consists of two discrete condensation stages. The first stage is conducted above the dewpoint of water, the condensate consisting primarily of light to heavy tar substances with boiling points well above 100°C. A second stage is used to remove excess water from the gas quench system.

The advantages of two-stage condensing over one-stage condensing are that byproduct tar residuals may be better reclaimed and, from an environmental point of view, that the second-stage byproduct wastewater tends to have lower quantities of entrained oils than does wastewater coming from a one-stage system quench. Two-stage cooling results in a wastewater much easier to treat for emulsified oil and tar removal than that from a one-stage condensate system.

As shown in Fig. 2.5, the spray cooler for the GFETC system is operated on a recirculating wastewater which is indirectly cooled in a heat exchanger. The blowdown from the decant tank of this loop is determined by the moisture content of the coal for high-moisture lignite coals and by water makeup rates in the case of bituminous coals.

Since evolution of the volatile matter occurs in the upper portions of the gasifier at temperatures much lower than those at which combustion and gasification occur, the organic pollutants from this type of facility are largely insensitive to the type of gasification which occurs below, i.e., whether it is slagging or dry ash.

The condensation of any excess steam introduced into a dry ash unit affects the level of effluent dilution. Therefore a distinction should be made by the designer between *quantity* of pollutants emitted per kg of coal gasified (preferably on a moisture- and ash-free — MAF — basis) and *concentration* ($mg\,l^{-1}$) which is often reported in the literature. The chemical nature of pollutants and the specific mix of substances in the wastewater treatment train are expected to be functions mainly of the parent coal for any fixed-bed gasifier when operated at similar pressures. This is clearly shown by the structures in Figs 2.1, 2.2 and 2.3.

Figure 2.6 is a block flow diagram of a possible Lurgi gasification step showing wastewaters and sludge production at various stages within the gas-producing complex. The operations included in this complex are those of coal preparation, gasification, hot gas cleaning and methanation, with auxiliary operations of the oxygen system, raw water plant, wastewater treatment plant and phenol and ammonia recovery facilities.

Figure 2.6 shows a separate stage of phenol removal prior to the wastewater treatment plant. Indeed, the Lurgi gasification plant package includes a solvent extraction step for phenol recovery. However, solvent extraction is economically viable only if a saleable product can be shown to exist.

Table 2.1 lists aqueous and solid waste streams expected to be generated at a proposed Lurgi gasification facility (Ghassemi *et al.*, 1979). It should be noted that these wastes include specific residuals from each of the areas outlined above.

2.3.3 Coal liquefaction facilities

The Solvent Refined Coal–II system (SRC–II) is one of the farthest advanced direct liquefaction processes. Preliminary engineering design and environmental impact statements have served as important sources of data for assessing environmental and health implications associated with full-scale facilities.

H-Coal, a second direct liquefaction plant considered in the USA, uses a catalyst to aid in liquefaction reactions. Determination of the presence or absence of potentially toxic substances formed by the spent catalyst is necessary. The US DOE supported a pilot plant at Catlettsburg, Kentucky, which uses Eastern bituminous coals as feedstock.

The Exxon Donor Solvent (EDS) system involves a non-catalytic liquefaction of coal in a hydrogen donor solvent with subsequent separation of solids and gases from the recirculating solvent liquid. This is followed by catalytic hydroprocessing of the liquids to provide regenerated donor solvent and hydrogenated end-products of improved quality (Sarna and O'Leary, 1979; Parker and Dykstra, 1978). The Exxon donor solvent process consists of the following operations:

(1) Coal feed and slurry preparation.
(2) Coal liquefaction–hydrogenation.
(3) Phase separation — gas and solids from the liquid product.
(4) Solvent hydrogenation and product upgrading, hydrotreating and fractionation.

The usual auxiliary and utility support operations are associated with a full-scale EDS facility:

- Steam power oxygen and hydrogen generation for use in the conversion process.
- Raw water, cooling and wastewater treatment and byproduct recovery of phenol, ammonia and sulphur if economically viable.

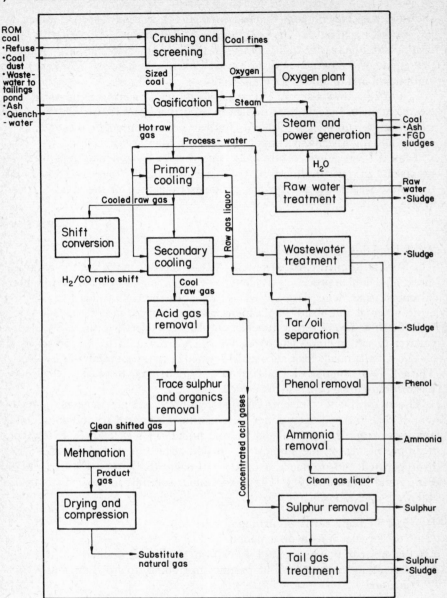

Fig. 2.6 Lurgi gasification (after Bern *et al.*, 1980)
● denotes residuals leaving the system

Table 2.1 Aqueous and solid waste streams generated at a proposed Lurgi gasification facility (after Ghassemi et al., 1979)

	Aqueous wastes	Solid wastes
GASIFICATION SYSTEM		
Coal preparation	Coal storage runoff	Coal fines, collected coal
	Tailings pond overflow	dust, coal refuse
Gasification	Ash quench slurry	Gasifier ash
Gas purification	Raw gas liquor	Spent shift catalyst
	Methanol/water still bottoms	Spent methanation guards
Gas upgrading	Methanation condensates	Spent methanation catalyst
	Drying and compression condensate	
WASTEWATER TREATMENT SYSTEM		
Process (gas liquor cleaning)	Phenol recovery — filter backwash	Phenol recovery — spent filter media
	Clean gas liquid	
Biological treatment		Biosludges
Process cooling	Cooling tower blowdown	
Plant runoff water treatment	Oil-free wastewater	Tarry/oily sludges
Ash slurry treatment	Clarified effluent	Dewatered ash solids
SULPHUR RECOVERY/AIR POLLUTION CONTROL		
Sulphur removal/recovery		Spent sorbents/reagents
		Sulphur
Tail gas desulphurization		Spent sorbents/reagents
		Sulphate/sulphite sludge
Utility flue gas cleaning		Same as above
UTILITIES		
Steam and power generation	Boiler blowdown	Boiler ash
Raw water treatment	Backwash water	Coagulation/settling sludges
		Regeneration blowdown sludges
Byproduct storage		Tarry/oily sludges

Figure 2.7 shows the H-Coal liquefaction system (Johnson et al., 1975), Fig. 2.8 the Solvent Refined Coal–II process (Cleland and Kingsbury, 1977) and Fig. 2.9 the Exxon Donor Solvent coal liquefaction system (after Parker and Dykstra, 1978). These flow diagrams serve to acquaint the environmental engineer with those individual process operations that generate wastewaters and other residuals from the various coal conversion processes.

2.4 Characteristics of coal conversion effluents

Little information was published until the late 1970s on composition and concentrations of coal conversion wastewater residuals. Most of the available

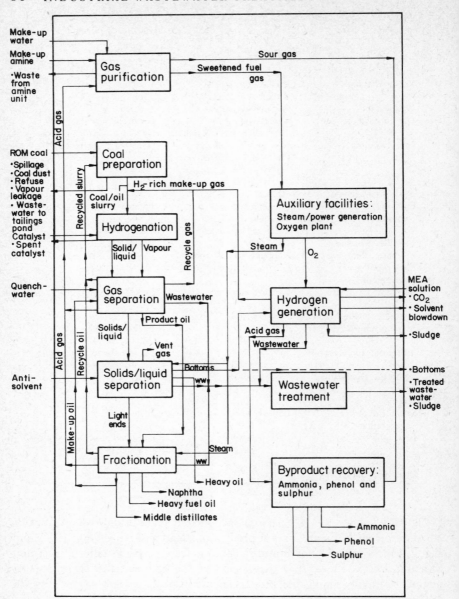

Fig. 2.7 The H–Coal liquefaction system (after Johnson *et al.*, 1975; Bern *et al.*, 1980)

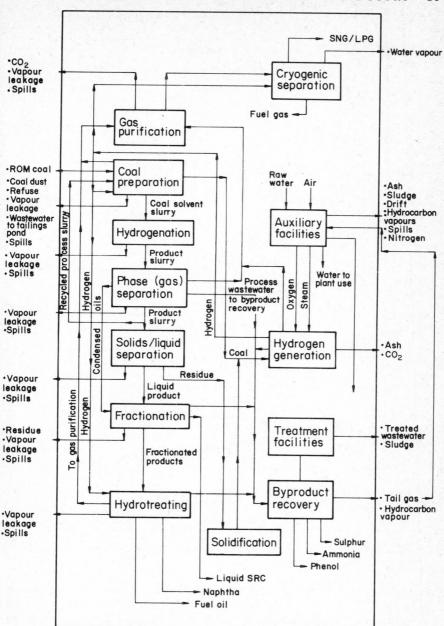

Fig. 2.8 The SRC–II process (after Cleland and Kingsbury, 1977; Bern *et al.*, 1980)

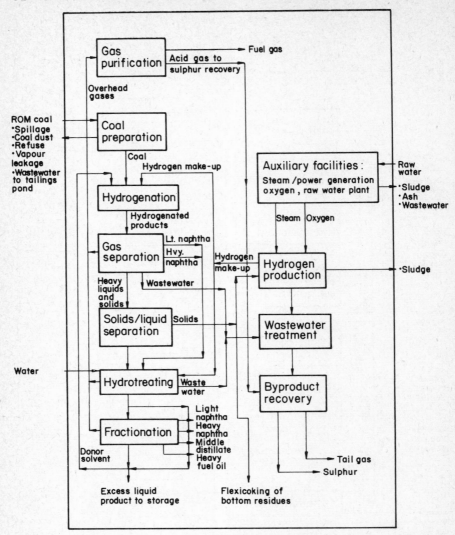

Fig. 2.9 The Exxon donor solvent coal liquefaction system (after Parker and Dykstra, 1978; Bern *et al.*, 1980)

data related to Lurgi-type fixed-bed processes since this type of gasification facility was the most common worldwide. Since 1977 there have been large expenditures of public funds by the US EPA, DOE and other US agencies on evaluation of the environmental impact of coal gasification, direct liquefaction and indirect liquefaction. The US EPA hired several large contractors to develop environmental studies of coal conversion. Included in the activities of these contractors was the onstream measurement of gaseous, liquid and solid residuals produced at coal conversion facilities in the USA and in other parts of the world. Chemical composition and estimated effects on health were assessed on the basis of such field analyses.

During the latter part of the 1970s and well into the 1980s funds were expended in developing second- and third-generation gasification facilities which promised to be more environmentally acceptable. In addition, industrial and university research within the USA began to examine alternative methods of treating liquid and solid effluents from coal conversion processes; effluents were characterized, analyzed and reported on a much broader basis than was previously available.

Several approaches may be employed to characterize the composition of coal conversion effluents. The simplest is to analyze for and report concentrations of major aqueous constituents such as BOD, COD, TOC, phenolics, ammonia, cyanides, thiocyanates, oil and tars, and salts. This approach may be used for regulatory purposes. Indeed, the 1977 'Best Practical Control Technology' as promulgated by the US EPA for the byproduct coke industry set discharge limitations only for BOD, total cyanide, phenol, ammonia, oil and grease, suspended solids and pH.

Published tables of composition of wastewater effluents taken from specific coal conversion facilities at specific times during the operation cycle are now available. However, information relating normalized pollution production parameters, i.e., kg of pollutant produced and m^3 of wastewater produced per kg of coal gasified (on a MAF basis), is not as plentiful since the development of this type of information involves a simultaneous knowledge of feedstock and overall facility operations.

Table 2.2 is a compendium of data illustrating normalized production of gasifier wastewater pollutants (kg pollutant produced per kg coal fed MAF). The processes listed include the fixed-bed Lurgi dry ash system at Westfield, Scotland, the Chapman gasifier gas producer at a US Department of Defense facility in Kingston, Tennessee, and the US DOE's GFETC slagging gasifier in Grand Forks, North Dakota. Also included in the table are other process designs: the Hygas process, developed by the Institute for Gas Technology (IGT), Chicago, Illinois; the GKT (formerly Koppers–Totzek) process; the Texaco process; and the Synthane fluidized bed process developed by the US DOE Pittsburgh Energy Technology Center in Pittsburgh, Pennsylvania. Pollution production parameters as a function of type of coal gasified are also reported. Where a range of data is known, the reference sources of data are given.

It is interesting to see that the Synthane process and the Lurgi dry ash process both produce large quantities of wastewater per unit of coal gasified MAF. The concentration of pollutants in such wastewaters tends to be low compared with, for example, pollutant concentrations associated with the US DOE GFETC slagging facility: comparison of phenol/water ratios, using lignite coals, shows virtually similar normalized phenol production rates but an

Table 2.2 Normalized production of gasifier wastewater pollutants (kg kg^{-1} coal MAF)

Parameter	FIXED-BED GASIFIERS					
	LURGI PROCESS — DRY ASH[a]				CHAPMAN GAS[b]	US DOE GFETC[c]
	High-volume bituminous		Sub-bituminous Montana	Lignite	Low sulphur metallurgical grade	Slagging gasifier lignite
	Illinois coal	Pittsburgh coal				
H$_2$O	2.2–2.7	2.96	1.48–1.51	1.99	0.24	0.312
BOD$_5$	1.3×10^{-2}	1.6×10^{-2}	$(2.0–3.3) \times 10^{-2}$		1.82×10^{-3}	$1.4 \times 10^{-2(d)}$
COD	$(2.3–3.2) \times 10^{-2}$	2.2×10^{-2}	$(3.1–5.2) \times 10^{-2}$		7.2×10^{-3}	$1.7 \times 10^{-2(d)}$
TOC					1.3×10^{-2}	$(5.0–7.0) \times 10^{-3}$
Phenolics	$(5–8) \times 10^{-3}$	6.4×10^{-3}	$(0.6–1.1) \times 10^{-2}$	5.7×10^{-3}	4.6×10^{-4}	7.3×10^{-3}
Organic acids	$(5.5–7.5) \times 10^{-3}$	1.7×10^{-3}	$(2.5–3.1) \times 10^{-2}$	4.7×10^{-4}		$(4.0–6.6) \times 10^{-3}$
Total ammonia	$(3.0–4.7) \times 10^{-2}$	4.2×10^{-2}	$(1.8–3.6) \times 10^{-2}$	6.0×10^{-3}	2.1×10^{-3}	
Total cyanide	3×10^{-5}	3.5×10^{-5}	$(5.1–7.8) \times 10^{-6}$	9.4×10^{-5}	3.2×10^{-7}	
Thiocyanate	$(3.4–4.3) \times 10^{-4}$	5.5×10^{-5}	$(0.09–4.0) \times 10^{-5}$	1.4×10^{-5}	1.5×10^{-6}	3.5×10^{-4}
Sulphide	1.1×10^{-3}	1.4×10^{-3}	$(0.08–2.8) \times 10^{-4}$	1.0×10^{-5}	1.3×10^{-3}	
Alkalinity as CO$_2$	$(1.8–2) \times 10^{-2}$	3.2×10^{-2}	$(2.1–2.9) \times 10^{-2}$	3.9×10^{-2}	1.8×10^{-4}	$(7.3–1.9) \times 10^{-2}$
Suspended oil and tar	$(4.7–5.7) \times 10^{-3}$	3.3×10^{-3}	$(1.0–1.6) \times 10^{-3}$		6.1×10^{-4}	3.2×10^{-4}
Total oil and tar			1.0×10^{-1}			
Total dissolved solids (TDS)	$(2.4–4.2) \times 10^{-3}$	3.7×10^{-3}	2.6×10^{-3}		4.5×10^{-3}	9.1×10^{-4}
Inorganic TDS	$(5.3–9.3) \times 10^{-5}$	3.6×10^{-4}	5.2×10^{-5}		7.0×10^{-5}	1.3×10^{-4}
Total suspended solids (TSS)					3.5×10^{-5}	1.5×10^{-7}
Chloride	$(2–3.6) \times 10^{-4}$	6.2×10^{-4}	$(3.9–5.9) \times 10^{-5}$			
pH	8.3–8.5	8.2	8.3			9.1–9.2

Table 2.2 cont'd

	OTHER GASIFIER CONFIGURATIONS					
	HYGAS[e]				KOPPERS–TOTZEK[f]	
Parameter	Bituminous	Sub-bituminous	Lignite	Bituminous	Sub-bituminous	Lignite
H_2O	0.98	1.14	1.1	0.35	0.33	0.31
BOD_5						
COD	9.9×10^{-4}	4.2×10^{-3}	5.5×10^{-3}	1.5×10^{-4}	1.4×10^{-4}	1.3×10^{-4}
TOC	1.2×10^{-3}	5.5×10^{-3}	2.9×10^{-3}	1.4×10^{-5}	1.3×10^{-5}	1.2×10^{-5}
Phenolics						
Organic acids						
Total ammonia	6.5×10^{-3}	7.6×10^{-3}	5.1×10^{-3}	3.0×10^{-3}	2.8×10^{-3}	2.5×10^{-3}
Total cyanide	2.5×10^{-6}	1.0×10^{-5}	1.4×10^{-6}	5.0×10^{-6}	4.9×10^{-6}	4.2×10^{-6}
Thiocyanate	9.9×10^{-5}	2.8×10^{-4}	4.8×10^{-4}	1.3×10^{-5}	1.2×10^{-5}	1.1×10^{-5}
Sulphide	7.7×10^{-4}	2.4×10^{-4}	1.5×10^{-4}	7.9×10^{-6}	7.8×10^{-6}	6.7×10^{-6}
Alkalinity as CO_2				8.0×10^{-3}	7.8×10^{-3}	6.7×10^{-3}
Suspended oil and tar	1.4×10^{-5}	8.8×10^{-5}	8.8×10^{-5}			
Total oil and tar						
Total dissolved solids (TDS)	1.1×10^{-3}	2.5×10^{-3}	2.4×10^{-3}			
Inorganic TDS						
Total suspended solids (TSS)	2.8×10^{-5}	1.2×10^{-3}	1.3×10^{-4}			
Chloride		7.4×10^{-6}				
pH	7.8	7.4	8.3			

Table 2.2 cont'd

Parameter	TEXACO[f] (Combined primary and secondary condensate)			SYNTHANE[g]		
	Bituminous	Sub-bituminous	Lignite	Bituminous	Sub-bituminous	Lignite
H_2O	0.08	1.0	1.4	2.0	1.4	1.6
BOD_5					1.4×10^{-2}	
COD				3.2×10^{-2}	2.8×10^{-2}	3.5×10^{-2}
TOC				1.1×10^{-2}	1.1×10^{-2}	2.0×10^{-2}
Phenolics				4.6×10^{-3}	4.4×10^{-3}	5.4×10^{-3}
Organic acids	6.0×10^{-4}	2.4×10^{-5}	2.3×10^{-6}			
Total ammonia	2.2×10^{-3}	3.9×10^{-3}	2.0×10^{-3}		7.0×10^{-3}	
Total cyanide				4.0×10^{-7}	1.0×10^{-7}	1.0×10^{-7}
Thiocyanate				3.4×10^{-4}	3.4×10^{-5}	2.7×10^{-5}
Sulphide					1.6×10^{-5}	
Alkalinity as CO_2	4.8×10^{-3}	8.1×10^{-3}	4.9×10^{-3}		8.7×10^{-4}	
Suspended oil and tar				9.8×10^{-4}	8.9×10^{-4}	
Total oil and tar				3.9×10^{-2}	1.5×10^{-2}	

[a] Ghassemi et al. (1979); Janes (1980); Neufeld et al. (1978).
[b] Iglar (1980).
[c] Ghassemi et al. (1978); Neufeld et al. (1978).
[d] Estimated from TOC.
[e] Ghassemi et al. (1978).
[f] Janes (1980).
[g] Neufeld et al. (1978).

eight-fold increase in concentration of phenolics (calculated as phenol/water) in the GFETC wastewater as compared with the Lurgi wastewater. Such increases in concentration of substances must be taken into account in the environmental engineering designer's assessment of the potential for phenolic recovery of different gasification systems. These variations in wastewater quantity, as indicated above, are due to moisture content of coal and, to a greater extent, quench-water and undecomposed steam originally fed to the gasification operation.

2.4.1 Composition by concentration

As with specific pollutant production parameters, available data on composition of coal gasification effluents lie in two general areas: data relating to generic pollutant concentrations, such as BOD, COD, TOC, etc., and data on the composition of specific organic constituent pollutants. This latter information has only been available in recent years with the use of more sophisticated methods of chemical analysis, such as gas chromatography, liquid chromatography and gas chromatography/mass spectrometry.

It should be pointed out that there is no 'typical composition' of wastewater from any given coal conversion process. The variables affecting wastewater composition are quite diverse, ranging from process characteristics and variations to temporal and wastewater analytical methodology used. Indeed, of a series of eight drums of wastewater collected within a 24-hour period of an operating pilot coal conversion facility, each was found to have a different composition with surprisingly wide variation from the first drum to the last, although all the drums had been filled with wastewater from the gasification unit presumably operating at the same steady-state condition. The environmental engineer must ensure that the wastewater treatment train that is proposed for a specific application is able to cope with wide variations of influent composition and flow rate.

Most of the published data consist of gross pollutant characterizations. The following are some examples of characterizations of a variety of coal conversion processes.

Table 2.3 compares reported process wastewaters from the Synthane fluidized-bed coal conversion facility, the Lurgi–Sasol facility, a Koppers–Totzek gasification process and the SRC–I process with typical coke plant weak ammonia liquor effluents (Hossain et al., 1979). This shows the Koppers–Totzek facility to produce relatively low levels of trace organics as compared with coke plant and coal conversion facilities. Janes (1980) reports characteristics of Lurgi coal conversion raw gas liquors as a function of coal type fed (Table 2.4). These values, reported in US EPA Draft (and finalized) Pollution Control Guidance Documents, are to be used in drawing up wastewater pollution control discharge permits. As can be seen from Table 2.4, the absolute and relative quantities of phenolics compared with total organics in Lurgi coal conversion facilities vary substantially, depending upon the nature of

Table 2.3 Comparison of process wastewaters

Parameter	Coke plant ammonia liquor ($mg\,l^{-1}$)	Synthane ($mg\,l^{-1}$)	Lurgi ($mg\,l^{-1}$)	Koppers–Totzek ($mg\,l^{-1}$)	SRC–I ($mg\,l^{-1}$)
pH	8.4	8.6	8.9	8.9	8.0
Suspended solids	4 000	600	5 000	50	300
Phenol	1 000	2 600	3 500		4 500
COD	10 000	15 000	12 500	70	15 000
Thiocyanate	1 000	152			
Cyanide	50			0.7	
Ammonia	5 000	8 100	11 200	25	5 600
Chloride	6 000	500		600	
Carbonate		6 000	10 000	1 200	
Sulphide	1 250	1 400			4 000

coal gasified. Examination of relative composition shows that, even if solvent extraction is used to remove the phenolic fraction, a substantial quantity of fatty acids and other organics still remains in solution.

Secondary conclusions may be derived from these data: for example, if one examines the COD equivalent of phenol, one can calculate a theoretical COD/phenol value of 2.38 by considering the chemical reaction

$$C_6H_5OH + 7O_2 = 6CO_2 + 3H_2O.$$

Thus, the COD equivalent (for Dunn County lignite coal) of the reported $2790\,mg\,l^{-1}$ of total phenols is equal to

$2790\,mg\,l^{-1}$ phenol $\times 2.38\,mg\,COD\,mg^{-1}$ phenol
$= 6649\,mg\,l^{-1}$ COD due to phenol alone.

This calculated COD level represents $6649/12\,500 = 53\%$ of the overall COD.

It thus appears that even if solvent extraction for phenolics were utilized for this waste, additional organics would remain in solution amounting to about 47% of the influent COD, which would have to be treated prior to discharge. Conventional techniques for treatment of such wastes are outlined below. As a point of comparison, Table 2.5 gives characteristics of the Koppers–Totzek condensate, also reported by Janes (1980). The data of Table 2.5 and Table 2.3 should be compared. As an example of substantial data discrepancies, reported alkalinity or carbonate–carbon concentrations and ammonia concentrations are substantially higher in Table 2.5 than in Table 2.3. The ammonia differences may be due to the lower operating temperatures used when lignite coal is gasified, thus minimizing ammonia decomposition in the hot gas phase.

Table 2.4 Characteristics of Lurgi raw gas liquors (Stream 51) (after Janes, 1980)

Constituent	Composition (in mg l^{-1} except pH)		
	Rosebud	Illinois No. 6	Dunn Co.
TDS	2 480	1 860	2 460
Sulphide (as H$_2$S)	55	290	49
Total sulphur (as S$^=$)	225	360	144
Thiocyanate	6	160	85
Cyanide (as HCN)	5	37	46
Carbonate (as CO$_2$)	13 600	7 780	10 000
Ammonia	7 610	4 800	2 577
Total phenols	5 000	2 280	2 790
Fatty acids	2 000	380	230
Tar and oil	150	500	300
SO$_3^=$/SO$_4^=$	500		150
TOC	7 640	2 980	4 190
Chloride	25	95	
BOD$_5$	10 600	3 600	5 600
COD	22 800	8 900	12 500
pH	8.2	7.8	8.9
Trace elements			
Ag	0.04	0.15	<0.2
As	0.4	0.5	3
B	14	50	3
Ba	<0.01	<0.1	
Be	0.59	0.6	0.008
Cd	0.26	<0.1	<0.3
Co	0.5	<0.1	<0.3
Cr	3.4	<0.1	<0.3
Cu	2.6	<0.1	0.3
F	50	42	0.4
Hg	0.15	0.6	<0.03
Li	2.5	<0.1	0.2
Mo	<0.01	<0.1	<0.2
Mn	<0.01	<0.1	
Ni	1.3	<0.1	<0.2
Pb	0.31	3	0.2
Sb	0.12	0.04	<0.03
Se	0.13	5	2
Sn	0.12	<0.1	<0.03
U	0.35	<0.1	<0.2
V	6.7	2	0.003
Zn	0.15	<0.1	0.4

2.4.2 Waste streams generated by synthesis operations in indirect liquefaction facilities

Catalytic methanation of synthetic gas is not expected to produce aqueous wastes containing dissolved organics species; indeed, these residuals may be used as boiler feed-water for in-plant use. Organic-based aqueous wastes, however, are expected to be generated by Fischer–Tropsch, the

Table 2.5 Characteristics of Koppers–Totzek condensate—lignite coal

Constituent	Concentration (mg l^{-1})
Carbonate CO_2	42 000
Sulphide	42
Cyanide	25
Thiocyanate	68
Ammonia	17 000
COD	420
TOC	40
$SO_3^=$	170
Non-volatile dissolved solids	<100

Mobile–M–gasoline and other indirect liquefaction synthesis operations. Janes (1980), in summarizing EPA contractor documents, reports that all waters produced during methanol synthesis will be found in the crude methanol product, and thus no aqueous waste is expected from this process unless refined methanol is the ultimate product. However, Janes (1980) does report the compositional properties of the Fischer–Tropsch process and the Mobile–M–gasoline process; this is shown in Table 2.6.

Table 2.6 Fischer–Tropsch and Mobile–M wastewater (after Janes, 1980)

Component	FT (mg l^{-1})	Mobile–M[a] (mg l^{-1})
BOD_5	13 000	11 000
COD	20 000	13 000
TOC	8 000	4 500
Organic acids	18 000	4 000 (mainly formic acid)
Ketones		4 000 (mainly acetone)
C_6^+ hydrocarbons		1 200 (mainly 'gasoline-type' isomers)

[a] More recent pilot data indicate that these values are high by $\sim \times 10$.

2.4.3 Specific constituent analysis

Singer et al. (1978) conducted an extensive literature search for specific constituents of coal conversion residuals. This was part of an overall research effort supported by the US EPA to assist in the development of a rational technological base for regulations applied to the synthetic fuels industry.

Table 2.7, reproduced from Singer's report, summarizes organic constituents found in coal conversion residuals. As can be seen, much of the available data comes from US government supported facilities and from analysis conducted on Lurgi-type facilities.

Bromel and Fleeker (1976) used chromatographic separations to examine specific organic species in raw and processed gasifier wastewaters from Lurgi–Sasol, South Africa, operations. They identified and quantified the specific

	Synthane	Oil shale	Syn-thane	COED	SRC	Lurgi-Westfield	Syn-thane	Lurgi-Sasol	Lurgi-type GFETC	Hydro-carboniz.	COED
MONOHYDRIC PHENOLS											
Phenol	1000–4480	10	2100	2100	✓	1200–3100	←2209	1250	5647		
o-Cresol	↑	30	670	650	✓	153–343	←⟶	340	↑		
m-Cresol	530–3580	←20	↑	↑	✓	170–422		360	1965		
p-Cresol	↓	↓	1800	1800	✓	160–302		290	←⟶		
2,6-Xylenol			40	30	✓						
3,5-Xylenol			230	240	✓			50			
2,3-Xylenol			30	40	✓				453		
2,5-Xylenol	140–1170		250	220	✓	100–393	2185				
3,4-Xylenol	↑		100	900	✓			120			
2,4-Xylenol	↓		—	—	✓				↓		
o-Ethylphenol			30	30	✓						
m-Ethylphenol					✓						
p-Ethylphenol					✓						
3-Methyl,6-Ethylphenol					✓						
2-Methyl,4-Ethylphenol	20–150				✓						
4-Methyl,2-Ethylphenol	←⟶				✓						
5-Methyl,1,3-Ethylphenol					✓						
2,3,5-Trimethylphenol					✓		66				
o-Isopropylphenol	✓						40				
DIHYDRIC PHENOLS											
Catechol	✓				↑	190–555				1700	✓
3-Methylcatechol					←⟶	30–394				↑	✓
4-Methylcatechol					↓	110–385				↓	✓
3,5-Dimethylcatechol						✓					
3,6-Dimethylcatechol	✓✓					0–45					
Methylpyrocatechol											
Resorcinol					✓	176–272				2000	✓
5-Methylresorcinol					✓	40–64				2000	✓
4-Methylresorcinol						0–36				2000	✓
2-Methylresorcinol	✓				✓					↓	
2,4-Dimethylresorcinol						✓					
Hydroquinone											

Compound				
Methylbenzoquinoline				
Tetrahydroquinoline				
Methyltetrahydroquinoline				
Isoquinoline	0–110			
Indole				
Methylindole				
Dimethylindole				
Benzoindole				
Methylbenzoindole				
Carbazole				63
Methylcarbazole				9
Acridine				
Methylacridine				4
ALIPHATIC ACIDS				
Acetic acid	600	620	600	171
Propanoic acid	210	60	90	26
n-Butanoic acid	130	20	40	13
2-Methylpropanoic acid	—	—	—	2
n-Pentanoic acid	200	10	30	12
3-Methylbutanoic acid	—	—	—	1
n-Hexanoic acid	250	20	30	1
n-Heptanoic acid	260	—	—	
n-Octanoic acid	250	—	—	
n-Nonanoic acid	100	—	—	
n-Decanoic acid	50	—	—	
OTHERS				
Benzofurans	10–110			
Benzofuranols	50–100			
Benzothiophenols	10–110			
Acetophenones	90–150			
Hydroxybenzaldehyde or benzoic acid	50–110			74

Table 2.7 cont'd

	Synthane	Oil shale	Syn-thane	COED	SRC	Lurgi-Westfield	Syn-thane	Lurgi-Sasol	Lurgi-type GFETC	Hydro-carboniz.	COED
POLYCYCLIC HYDROXY COMPOUNDS											
γ-Naphthol	←——		10								
β-Naphthol	30–290		30								
Methylnaphthol	20–110										
Indenol	←——→										
C₁-Indenol	40–150										
4-Indanol	←——→						66				
C₁-Indanol	0–110										
Biphenol											
Biphenyl										19	
MONOCYCLIC N-AROMATICS											
Pyridine								117			
Hydroxypyridine										10	
Methylhydroxypyridine										10	
Methylpyridine	←——— 30–580 ———→						⎫	104		←	
Dimethylpyridine							⎬ 5	<1		20	
Ethylpyridine							⎭			↙	
C₃-Pyridine											
C₄-Pyridine											
Aniline							21	12			
Methylaniline							9				
Dimethylaniline							11				
POLYCYCLIC N-AROMATICS											
Quinoline	←— 0–100 —→				⎫		7				
Methylquinoline					⎬		27				
Dimethylquinoline					⎬						
Ethylquinoline					⎬						
Benzoquinoline					⎭						

Table 2.8 Concentration of organic compounds and their equivalent COD and TOC values (mg l^{-1}) for the separated and clean Lurgi gas liquor at Sasol, South Africa (Ghassemi et al., 1979)

Compound	Separated gas liquor mg l^{-1}	COD	TOC	Cleaned gas liquor mg l^{-1}	COD	TOC
FATTY ACIDS						
Acetic acid	171	183	68.4	123	131.6	49.2
Propanoic acid	26	39.3	12.7	30	45.3	14.7
Butanoic acid	13	23.7	7.8	16	29.1	9.6
2-Methylpropanoic acid	2	3.8	1.1	5	9.5	2.7
Pentanoic acid	12	24.5	7.1	7	14.3	4.1
3-Methylbutanoic acid	1	2.1	0.6	5	10.5	2.9
Hexanoic acid	1	2.2	0.6	8	17.7	5.0
	226	278.6	98.3	194	258	88.2
MONOHYDRIC PHENOLS						
Phenol	1250	2975	963	3.2	7.6	2.5
2-Methylphenol	340	857	265	<0.2	<0.5	<0.2
3-Methylphenol	360	907	277	<0.2	<0.5	<0.2
4-Methylphenol	290	731	226	<0.2	<0.5	<0.2
2,4-Dimethylphenol	120	314	95	NF		
3,5-Dimethylphenol	<50	<131	<39.5	NF		
	2410	5915	1866	3.2	9.1	3.1
AROMATIC AMINES						
Pyridine	117	261	88.9	0.45	1.0	0.34
2-Methylpyridine	70	169	53.9	<0.05	<0.12	<0.04
3-Methylpyridine	26	62.7	20.0	<0.05	<0.12	<0.04
4-Methylpyridine	6	14.5	4.6	<0.05	<0.12	<0.04
2,4-Dimethylpyridine	<1	<2.5	<0.8			
2,5-Dimethylpyridine	<1	<2.5	<0.8			
2,6-Dimethylpyridine	<1	<2.5	<0.8			
Aniline	12	28.9	9.2			
	231	544	179	0.45	1.4	0.5
Total	2867	6738	2143	198	269	92

wastewater organic constituents shown in Table 2.8, of which phenol and its methylated substituents of cresols (methylphenols) and xylenols (dimethylphenols) were major organic components. Other major classes identified in this wastewater were fatty aliphatic acids and aromatic amines consisting of heterocyclic pyridine and its methyl derivatives. Table 2.8 also indicates concentrations of aqueous constituents after solvent extraction at the Lurgi–Sasol facility. It is apparent that extraction removed most of the monohydric phenolics, comprising phenol and methyl and dimethyl derivatives. Fatty acid constituents of acetic and larger organic acids were not removed in the solvent extraction step. Accordingly, these data indicate that solvent extraction may meet only a portion of the overall wastewater treatment needs, subsequent processing being necessary for removal of other organics, thiocyanates and ammonia-containing substances. Table 2.9 summarizes Lurgi–Westfield data and shows that a substantial fraction of the phenolics in Lurgi coal conversion wastewaters are dihydric phenolics — substances which are less readily extracted in a solvent extraction procedure.

Table 2.10 is a gas chromatography analysis of Synthane wastewaters from the gasification of Illinois coals (Neufeld and Spinola, 1978). These data show that the bulk of the specific organic constituents in Synthane wastewaters are phenols and o-cresol with almost an equal contribution by m- and p-cresol, and 2,4- and 2,5-xylenol. Such substances are fragments of the theoretical coal molecule shown in Fig. 2.2 and may be considered as being volatilized during the gasification process.

Table 2.9 Lurgi–Scotland coal gasification wastewater components (concentrations in $mg\,l^{-1}$) (Bertrand et al., 1974)

	Oil water	Tar water
Ammonia, total (NH_3)	9597	1795
Carbonate ($CO_3^=$)	17655	1128
Cyanide (CN^-)	2.6	7.8
Chloride (Cl^-)	11.3	4.3
Dihydric phenols	1869	2917
Fatty acids	228	696
Ferrocyanide ($Fe(CN)_6$)	10.5	4.2
Extractable by ether	100–500	1000–5000
Monohydric phenols	3178	2864
Sulphate ($SO_4^=$)	74.1	90.6
Sulphide ($S^=$)	0.77	0.7
Thiocyanate (CNS^-)	41.2	nil
Thiosulphate (S_2O_3)	15.8	9.0
Total alkalinity (as $CaCO_3$)	5000	2500
Suspended solids	340	100
pH	8.0	9.4
Total iron (Fe)	2.0	2.3
Calcium hardness (as $CaCO_3$)	nil	nil
Magnesium hardness	nil	nil
Temperature	21°C	71°C

Table 2.10 Gas chromatography analysis of Synthane process wastewater from Illinois No. 6 coal (Neufeld and Spinola, 1978)

	Concentrations as received (mg l^{-1})
Unidentified low-boiling compounds (possibly acetone, methanol, benzene, etc.)	190
Pyridine and picolines	18
Lutidines	5
Naphthalene and aniline	21
Toluidines	9
2-Methylnaphthalene and xylidine	11
2,6-Xylenol	24
Quinoline	7
Phenol and o-cresol	2209
m- and p-cresol, 2,4- and 2,5-xylenol	1626
Methylquinoline	27
2,3-Xylenol	50
3,5-Xylenol, m- and p-ethylphenol	366
3,4-Xylenol	119
3-Ethyl-5-methylphenol	66
C_3-substituted phenol	40
4-Indanol	66
Indole	63

Table 2.11 Concentration (mg l^{-1}) of phenols in tar liquor and oil liquor at Westfield Works, February 1976 (after Sinor, 1977)

	Tar liquor	Oil liquor
PHENOLS (total)	3570	5100
Monohydric phenols	1843	4560
Phenol	1260	3100
o-Cresol	155	343
m-Cresol	170	422
p-Cresol	160	302
Total xylenols	100	393
Monohydric phenols as percentage of total phenols	52%	89%
OTHER PHENOLS		
Catechol	555	190
3-Methyl catechol	394	80
4-Methyl catechol	385	110
3,5-Dimethyl catechol	trace	trace
3,6-Dimethyl catechol	45	trace
Resorcinol	272	176
5-Methyl resorcinol	40	64
4-Methyl resorcinol	36	
2,4-Dimethyl resorcinol	trace	trace

There is little information on the distribution of various phenolics such as monohydric and polyhydric phenolics in the Lurgi and other fixed-bed coal conversion streams. Table 2.11 gives a distribution of phenols and phenolics in tarry liquors and oily liquors from the Westfield, Scotland, coal conversion facility (Sinor, 1977). This table shows that monohydric phenols are approximately 52% of total phenols in tar liquors and approximately 90% of total phenols in oil liquors. Other phenolics of the cresol, catechol and resorcinol families are present in Lurgi facility wastewaters, as may be expected from a closer examination of the model coal molecule. Data measured on existing coal conversion facilities of the Lurgi type are given in Table 2.8 for Sasolburg, South Africa, and in Tables 2.9 and 2.11 for Westfield, Scotland.

Table 2.12 lists contaminants from a variety of coals gasified in the US DOE Pittsburgh Energy Technology Center's pilot Synthane fluidized-bed coal conversion facility. As can be seen, the magnitude of specific constituents varies with coal type; however, the relative distribution of phenolics is approximately uniform. The constituents reported by Schmidt et al. (1974) are also somewhat similar in a qualitative manner to the data outlined by Bromel and Fleeker (1976) for the Lurgi–Sasol coal conversion facilities.

Not all coal conversion wastewaters contain the various species of phenolics and other trace organics indicated in Tables 2.2–12. While it may be predicted that the composition of wastewater from coal conversion facilities is dependent upon specific process conditions, such as operating temperature, pressure, mode of contact between coal and steam, gas clean-up, etc., it appears, particularly in the light of Table 2.2, that the dominant factor is the oxygen/coal ratio and mode of contact. For example, Koppers–Totzek and GKT gasifiers have been shown to produce relatively low levels of total organics. This is because flame temperatures in the gasifier section are high, so that any phenolics formed are oxidized. Fixed-bed facilities, such as the Lurgi and GFETC facilities, have a zone of devolatilization which permits a broad spectrum of contaminants to be produced. Of the category 'phenolic producer gasifiers', it is interesting to note that the composition of coal gasification and liquefaction wastewaters appears to be relatively uniform, particularly with respect to phenolic constituents. Less information, however, is available regarding the presence or absence of specific nitrogen-containing aromatics, heterocyclic compounds and polynuclear aromatic hydrocarbons.

Stamoudis et al. (1979, 1980) at the US Argonne National Laboratory provide some insight into the semi-quantitative composition of coal conversion residuals. The Argonne National Laboratory conducted acid, base and neutral solvent extractions of samples of Hygas wastewaters and GFETC slagging fixed-bed coal gasification wastewaters. As may be expected, the acid fraction of the effluent contained primarily phenol and single ring alkylated phenolic compounds with phenols and cresols constituting the largest fraction of observed organics. The base fraction consisted primarily of nitrogen heterocyclic compounds including pyridines, quinolines and indoles in their alkylated

Table 2.12 Contaminants in product water from Synthane gasification of various coals (concentrations in mg l^{-1}) (after Schmidt et al., 1974)

Component	Illinois No. 6 (HVBB)	Montana (Sub.)	North Dakota (Lig.)	Wyoming (Sub.)	West Kentucky (HVBB)	Pittsburgh (HVAB)
Phenol	3400	3160	2790	4050	2040	1880
Cresols	2840	870	1730	2090	1910	2000
C$_2$-Phenols	1090	240	450	440	620	760
C$_3$-Phenols	110	30	60	50	60	130
Dihydrics	250	130	70	530	280	130
Benzofuranols	70	80	60	100	50	70
Indanols	150	140	110	110	90	120
Acetophenones	100	—	—	60	50	—
Hydroxybenzaldehyde	60	—	—	—	—	—
Benzoic acid	110	160	140	80	160	170
Naphthols	160	70	50	60	80	20
Indenols	90	10	10	—	—	110
Benzofurans	—	—	—	—	—	—
Dibenzofurans	—	—	—	40	20	60
Biphenols	20	—	10	20	70	20
Benzothiophenols	40	—	220	120	30	540
Pyridines	60	270	10	—	—	—
Quinolines	110	20	30	20	40	10
Indoles	—	70				40

and methylated forms. The neutral fractions consisted primarily of toluene and alkylated cycloalkanes, cycloalkenes and polycyclic aromatic hydrocarbons.

2.5 Effluent standards

2.5.1 US guidelines for comparable industries

At present, the USA does not have guidelines for effluent discharges from coal conversion facilities. Effluent discharge permits (air, water, solid waste) for the first few US-based plants will be developed on a case-by-case basis. A coal conversion facility, or any other facility that is constructed with US government funds, is required by law to issue an 'environmental impact statement'. Usually the cognizant federal agency is responsible for preparing the environmental impact assessment statement, which must go through a public hearing before it is accepted. Private developers, however, may not be required to produce environmental impact statements prior to construction.

The local state governments are responsible for administering water pollution control laws. Individual states are responsible for granting discharge permits for point sources within their boundaries. These regulations are expected to apply equally to coal conversion facilities.

There is a growing trend on the part of the US EPA to develop effluent *guidelines* for various industrial categories. These guidelines, issued in accordance with the Water Pollution Control Act of the US Congress, outline technologies designated as 'Best Practical Treatment' (BPT), a level of achievement originally proposed for 1977 discharges, and 'Best Available Technology' (BAT), proposed as a level of achievement for the early 1980s. It should be noted that the recommendations for BPT and BAT are guidelines only. In practice, however, individual states utilize such guidelines and adhere to them closely when granting permits. The industry most similar to the coal conversion industry for which BAT guidelines have been developed is the byproduct coking industry, considered to be an integral part of the iron and steel industry. Table 2.13 gives the effluent limitation guidelines for the byproduct coking industry, set out in the 1981 Federal Register by the US EPA and summarized by Greenfield and Neufeld (1981). The BAT effluent limitations were based upon a comprehensive study by the US EPA and its contractors of the status of wastewater treatment within the iron and steel industry. The EPA has published a discussion of technologies applied by the iron and steel industry (EPA, 1980).

Table 2.13 outlines both BPT and BAT effluent limitations for the coke subcategory of the iron and steel industry. Similar guidelines may well become the basis for regulations for the synthetic fuels industry. It is important to note that the guidelines for BPT include those for gross pollutants such as total suspended solids (TSS), oil and grease, ammonia nitrogen, cyanide, phenol and pH. The permitted concentration of ammonia nitrogen is similar to that which

Table 2.13 Effluent limitation guideline for byproduct coke subcategory of the iron and steel industry (after Greenfield and Neufeld, 1981)

	Maximum for any one day per calendar month		Monthly average	
Pollutant parameter	Discharge load (kg per 1000 kg)	Discharge concentration[a] (mg l^{-1})	Discharge load (kg per 1000 kg)	Discharge concentration[a] (mg l^{-1})
BPT EFFLUENT LIMITATIONS				
TSS	0.225 0	359.7	0.075 0	119.9
Oil and grease	0.032 7	52.3	0.010 9	17.4
Ammonia–N	0.273 6	437.4	0.091 2	145.8
Cyanide	0.065 7	105.0	0.021 9	35.0
Phenol	0.004 5	7.2	0.001 5	2.4
pH—within the range pH 6.0–9.0				
BAT EFFLUENT LIMITATIONS				
Ammonia–N	0.051 10	81.7	0.009 57	15.3
Cyanide	0.003 20	5.1	0.001 60	2.6
Phenol	0.000 0640	0.10	0.000 0160	0.026
Benzene	0.000 0638	0.10	0.000 0319	0.05
Naphthalene	0.000 0128	0.02	0.000 0064	0.01
Benzo(a)pyrene	0.000 0256	0.04	0.000 0128	0.02

[a] Concentrations are based on effluent flows of 0.63 m^3 per tonne of coke for BPT and BAT.

is expected in the effluent from a properly operated free and fixed ammonia still. The levels permitted for phenol are typical of those levels in effluents from conventional dephenolyzers. The BAT limitations are much more stringent. The BAT average ammonia discharge concentration is approximately 15 mg l^{-1}, a value which is achievable commercially only by means of alkaline chlorination ion exchange or by the use of biological nitrification in a one-stage or two-stage activated sludge process.

The regulations published by the US EPA in 1981 are based upon pollution production parameters, kg pollutant discharged per tonne of coke manufactured. This is a substantial departure from the general concept of effluent concentration limitations. Compliance appears to involve detailed records of concentrations, flows and product production of the overall plant.

A second substantial deviation from past practice was made by the US EPA in developing effluent guidelines that may be applicable to the emerging synthetic fuels industry. The US EPA has introduced the concept of surrogate pollutants for regulatory purposes for the coke plant industry. It was acknowledged by the US EPA that a requirement to monitor for all of the 'priority pollutants' would pose an undue burden on industry. Accordingly, the US EPA regards phenol, measured by the 4-aminoantipyrine (4-AA) technique outlined in *Standard Methods* (APHA, 1980), as indicative of the presence or absence of acid-extractable priority pollutants, benzene as indicative of the volatile fraction, and naphthalene and benzo(*a*)pyrene as indicative of the presence or absence of the base and neutral portions of those 'priority pollutants' that are present in coke plant wastewaters. The concept of surrogate pollutants has not yet been applied to the coal conversion industry, but its introduction in the US iron and steel industry indicates the direction that regulations may take for the coal conversion industry.

2.5.2 Environmental impact — data and technological development

In 1981, the US Congress formed the Synthetic Fuels Corporation with responsibility to provide federal loan assistance for the construction of commercial-scale coal conversion facilities, with the possibility of future price guarantees, if necessary, for the resulting product. Many proposals have been put forward to the Synthetic Fuels Corporation. In addition to government-supported coal conversion facilities, there exists a number of privately developed facilities that have refused US government assistance. Private development is proceeding in the areas of coal gasification, coal liquefaction using indirect liquefaction processes, and combined cycle gasification facilities where the product gas is utilized directly in a turbine. Much of the data and status of such privately developed facilities is currently not reported and is generally regarded as confidential.

One example of a comprehensive environmental technology plan and environmental impact statement is the *Final Environmental Impact Statement*,

Volumes I and II, for the Solvent Refined Coal–II Demonstration Project, proposed for Fort Martin, West Virginia. This document, issued by the US Department of Energy in January 1981, relates to a proposal by the USA, the Federal Republic of Germany, Japan and SRC–II International Inc. to build and operate a SRC–II demonstration plant on a cost-shared basis. This project has been designed to demonstrate the technical operability, economic viability and environmental acceptability of the SRC–II process. One of its primary objectives is to obtain an environmental database. The data are to be used by regulatory agencies in setting standards for acceptable environmental quality and by the industry to evaluate the economics of a commercial facility that includes required environmental controls. It is assumed by the US DOE and their partners that any decision to construct a commercial-scale facility will incorporate information gained from the construction and operation of the demonstration facility in an environmentally acceptable manner. It is significant that one of the principal goals of demonstration plants, as outlined in the Environmental Impact Statement, is to develop environmental protection information that would be applicable both to regulation development and to assessment of technological viability. However, the SRC-II project has not been constructed.

2.6 Design of coal conversion wastewater treatment systems

2.6.1 Principles and treatment methods

From the discussion outlined above, the magnitude of flow from a commercial coal conversion facility would be about equal to that expected from a small-to-medium-sized municipality, i.e., in the range $3800-38\,000\,\text{m}^3\,\text{d}^{-1}$ (a population equivalent in the range 10 000 to 100 000). Coal refinery processes may be categorized as those that give relatively high organic loadings and those that give relatively low or negligible organic loadings. A feature common to most condensate waters is the substantial concentration of dissolved ammonia, which results from the high partial pressures of ammonia in the uppermost reaches of the gasifier. High-temperature operation can cause decomposition of ammonia to nitrogen. Very high levels of dissolved ammonia and carbon dioxide indicate that process condensates are strongly buffered and neutralization of raw gasifier waters would require extensive quantities of acid or base chemicals. Typical pH values for process condensates are in the range of about pH 8 to 9.

The environmental process design engineer is required to develop combinations of unit operations that can treat coal conversion effluents and residuals to levels consistent with regulatory constraints, and in a fashion that is both economical for the plant owner and acceptable to the increasing scrutiny of the public sector. The approaches that may be utilized include:

- physicochemical treatment followed by broad-brush biological treatment, the treated effluent being either recycled to the plant complex or discharged to surface waters;

- processing techniques involving limited physicochemical processing, with an emphasis on evaporative procedures for disposal; such evaporative procedures include discharge of partially treated wastewaters into cooling towers and incorporation of multiple-effect evaporation for volume reduction, with ultimate disposal of concentrated residuals on the land.

While this latter approach satisfies the concept of 'zero liquid discharge', it should be recognized that other discharges to air and land may result in unacceptable environmental impact.

Research is under way, supported by the US DOE and the US EPA, at various industrial and university centres with the object of developing environmentally acceptable control strategies and technologies for the emerging coal conversion industry. Research leading to rational design parameters is essential if the coal conversion industry is to achieve commercial status. As an example, research at the University of Pittsburgh aims to develop a combination of wastewater treatment unit operations for the treatment of fixed-bed coal conversion wastewaters and to modify such units to render sludge residuals non-toxic and non-mutagenic. Researchers in this and in other major university centres are examining the fates of trace organics through treatment processes and are evaluating the impacts of non-biodegradable residuals from synthetic coal conversion effluents.

Biological processes are usually considered a viable option because of their wide applicability to a variety of wastes and their long history of commercial operation. Biological processes, however, are limited in terms of their ability to be applied directly to raw coal conversion wastewaters which may contain high levels of ammonia and other potentially inhibitory substances. Accordingly, as in coking facilities, pretreatment of coal conversion effluents is required prior to biological oxidation. If concentrations of key substances are sufficiently high, profitable product recovery may be possible. Typical substances that are often recoverable include ammonia and phenolics — one of the earlier uses of coal gasification was to produce ammonia for use in fertilizer plants. Phenol recovery is economically viable only if phenol levels are sufficiently high and product impurities are not critical.

Since biological processes are living systems, temporal deviations in process efficiency and process stability are often observed. In industrial applications, such deviations often stem from changes in the nature of feed and ambient conditions of the biological reactor. Physicochemical operations, however, may accommodate substantial changes in the nature of influent wastewater by means of appropriate instrumentation and sensing devices. Unfortunately, the present state of art for biological systems precludes such sensing and feedback loops.

However, research is under way which may revolutionize biosystem process stability and applicability within the next decade.

The overall goal for pretreatment in coal conversion comprises the removal of oil and grease, sulphide oxidation or stripping, ammonia removal to levels compatible with biological systems, cyanide removal, pH adjustment as required, and removal, if possible, of potentially inhibitory and otherwise non-biodegradable residuals.

Biological processing techniques have as their goal the effective and consistent removal of biodegradable organic constituents. As listed in Tables 2.2–12, the biodegradable residuals in coal conversion wastewaters include the family of fatty acids, simple and some complex phenols, thiocyanates, simple cyanides, and other BOD-exerting materials. As outlined by McKenna and Heath (1976), polynuclear aromatic substances may be biodegraded, albeit slowly, by certain wastewater micro-organisms. These authors delineated the structural limits of biodegradability of polynuclear aromatic substances and measured the persistence of selected PNAs even after degradation by mixed microbial cultures. They found that the number of fused rings, the size, position and number of ring substituents, and degree of saturation influenced the initial rates of aromatic oxidation. From a process environmental engineering point of view, biological systems operated at long sludge ages may have the best capability to oxidize the polycyclic aromatics that may be found in coal refinery effluents.

Two major approaches may be taken to biological degradation of organic substances: aerobic systems and anaerobic systems. The feasibility of using mesophilic anaerobic fermentation for the treatment of petrochemical wastes as an alternative to conventional aerobic biological systems was examined by Hovious *et al.* (1972). Neufeld *et al.* (1980) found that anaerobic phenol degradation biokinetics followed a substrate inhibition model whereby a maximum of about 700 mg l^{-1} could be degraded anaerobically before the onset of inhibition and thus unstable conditions. High levels of phenol were found to decompose into simpler straight-chain volatile acids; owing to high levels of phenolics and organic acids in solution, no methane production was observed. Their data show that the maximum specific rate of anaerobic phenol degradation is only about 0.5–0.6% of the aerobic specific rate of phenol utilization. These kinetics may explain the anaerobic degradation of phenolics in stream sediments under natural conditions; such kinetic evaluations also explain upper technological limitations for the use of anaerobic processes for wastewaters with high phenolic content. Biokinetic coefficients and sludge yield coefficients were also presented.

Because of their relatively high biokinetic rate, aerobic processes of both fixed film (trickling filters and rotating biological contactors) and suspended film (activated sludge type and its modifications) have been proposed for organic oxidation of pretreated coal conversion wastewaters. A completely mixed reactor design is considered preferable to plug-flow-type designs, since it provides for dilution within the reactor of any shock loading to the biological system. It is recognized that a kinetic penalty must be paid for incorporation of completely mixed systems. However, the benefits to be derived in terms of process stability should far outweigh the inherent disadvantages. Of course, it is the responsibility of the environmental designer to ensure that the pretreatment train produces an acceptable feed to the biological unit.

One of the earliest holistic evaluations of treatment of coal conversion demonstration and pilot plant wastewaters was reported by Johnson et al. (1977). Effluents from the Synthane fluidized-bed coal gasification facilities were used to develop gross characterizations, and preliminary inquiries into the areas of oil and tar removal, and ammonia and trace organic removal. Neufeld et al. (1978) developed the biokinetics necessary for the design of activated sludge wastewater treatment plants utilizing Synthane effluents. It was found that dilutions up to 30% wastewater in tap water were capable of providing stable operations for the Synthane application. It was also found that approximately 15% of the wastewater-soluble carbon influent to the biological treatment process, measured as TOC (soluble total organic carbon), appeared to be non-biodegradable and passed through the biological reactors. Neufeld and Spinola (1978) applied ozone to samples of coal conversion effluents and presented data relating to reformation of some of the key organic species present.

Cooke and Graham (1965) present data on the biological treatment of Lurgi plant wastewaters. Sack (1979) conducted studies with fixed-bed coal gasification wastewater from the Morgantown Energy Technology Center. His studies showed that high aqueous salt concentrations, as derived from utilization of caustic for pH adjustment in ammonia stripping, resulted in severe inhibition of downstream biological reactors and thus required both long sludge ages and extraordinarily long detention times to obtain suitable phenolic removal.

Reap et al. (1979) present data on wastewater characteristics and proposed treatment technology for the H-Coal liquefaction process. Luthy and Tallon (1978) and Luthy et al. (1979) present laboratory-based experimentation on the biological treatment of Hygas pilot plant wastewaters and fixed-bed coal gasification wastewaters.

Singer (1979) and Singer et al. (1979) conducted long-term biological treatability experiments on a laboratory-developed synthetic mixture simulating coal gasification wastewater on the basis of literature reviews of reported coal gasification effluents. The significance of the work by Singer et al. (1979) is that the fate of key specific trace organics could be monitored through the biological treatment step since controls were based on influent composition. Questions are raised, however, as to the direct applicability of the results of this research approach to commercial-scale and demonstration-scale operation. Drummond et al. (1980) report upon biokinetic evaluation of both coal gasification and coal liquefaction wastewaters in research conducted at the US DOE Pittsburgh Energy Technology Center.

Building upon the results of prior investigations, Neufeld et al. (1981c; 1981d) developed a series of pretreatment procedures that enabled wastewater derived from a stirred fixed-bed gasification process, modified from the Wellman–Galusha design, to be processed in a stable fashion in an aerobic activated sludge biological reactor. This study represented the first time sufficient quantities of field-produced wastewater were provided to allow both modification of pretreatment trains and modification of the biological reactor system. Earlier research used relatively limited quantities of wastewaters (600–1500 litres), whereas this study was conducted over a long term using approximately 5700 litres and 8000 litres of wastewater, respectively, in the first and the second phases of research. The limited quantities of wastewater used in earlier investigations restricted the amount of data available for the examination of a range of independent variables investigated for the pretreatment scheme and postbiological treatment scheme.

2.6.2 Treatment train design: examples (for medium level phenolic wastes)

Table 2.14 characterizes samples from three shipments of coal conversion wastewater from the US DOE METC stirred fixed-bed coal gasification facility. Sample No. 1 was an old sample of wastewater collected at least one year prior to analysis. Samples Nos 2 and 3 were from two consecutive gasification runs. The aged sample is seen to contain high levels of COD, and relatively lower levels of phenol. The phenol/TOC ratio for this particular run was 0.069, a value considerably lower than that reported as existing in other coal conversion facilities. Table 2.15 gives phenol/TOC ratios for various coal conversion processes. Note for comparison purposes that the phenol/TOC ratio for molecular phenol (C_6H_5OH) is 1.3. Comparison of the 'total residues'

Table 2.14 Characterization of three shipments of wastewater from the US DOE METC fixed-bed coal gasification facility (concentrations in $mg\,l^{-1}$)

	Sample No. 1 (aged sample)	Sample No. 2 (Run 94)	Sample No. 3 (Run 95)
Phenol	970	2 375	3 750
COD	53 024	13 350	12 750
TOC	14 102	5 374	5 390
TIC	30	242	4 350
Total residue	72 334	1 349	4 420
Fixed residue	1 334	143.2	630
Volatile residue	71 000	1 205.8	3 790
Freon sol. oil and grease	356	1 495.0	—
Acetone sol. oil and grease	1 633	106.3	—
pH	7.5	8.0	8.8
Alkalinity (pH 4.5)	2 100	23 750	21 855
Thiocyanate	—	372	—
Ammonia	11 000	3 200	7 000
Phenol/TOC ratio	0.069	0.44	0.70

Table 2.15 Phenol/TOC ratios for various coal conversion processes (after Neufeld et al., 1981c)

Coal type	Process	Phenol/TOC (kg kg^{-1})
Lignite	Synthane–PDU	0.265
Lignite	GFETC–400 psi	0.819
Lignite	Hygas	0.292
Sub-bituminous	Synthane	0.395
Bituminous	Synthane	0.432
—	METC	0.44–0.70

and the 'volatile residues' values indicates that the aged sample (No. 1) is considerably 'dirtier' than samples Nos 2 and 3; however, this sample is significantly less alkaline than the other two samples. Samples Nos 2 and 3 have phenol/TOC ratios in a typical range, with ammonia NH$_3$ levels varying from 3200 mg l^{-1} to 7000 mg l^{-1}. It should be noted that in all three shipments the level of phenols was less than what might be considered economically recoverable by solvent extraction.

Figure 2.10 shows the wastewater treatment train developed at the University of Pittsburgh to treat fixed-bed coal gasification wastewaters. The processing train consisted of a combination of steps as follows:

Step 1. Air or steam stripping. This was accomplished in the laboratory via aeration or steam sparging of raw wastewaters. Such aeration liberates noticeable quantities of hydrogen sulphide, free cyanides, volatile organic fractions and free ammonia.

Step 2. Lime addition. Slaked lime was added to wastewater in sufficient quantities to bring the pH to a range of 10.5 to 11. A mixing time of about 30 minutes was utilized to ensure completion of any reactions. It should be noted that in commercial-sized systems a two-stage ammonia stripping operation would be included, consisting of air or stream stripping of the 'free' ammonia followed by a pH adjustment to 10.5 to 11 for stripping of the 'fixed' ammonia fraction. The advantage of such two-stage operations is that the first stage reduces wastewater alkalinity and thus reduces the quantity of lime required for ammonia stripping. Proprietary ammonia recovery systems that are available commercially may be applicable to synthetic fuel wastewaters but the economics and suitability should be evaluated for the individual application proposed.

Step 3. Sludge removal. Sludge formed from lime reaction with gasifier wastewater was removed just prior to ammonia stripping. The colour of the resultant sludge was brown, suggesting the presence of organic materials. Such sludges were bioactive as indicated by Ames test in the laboratory. Despite the excess lime used in this step, the ratio of volatile organic solids to total solids (a measure of organic content) of the sludge was in the range 10% to 13%.

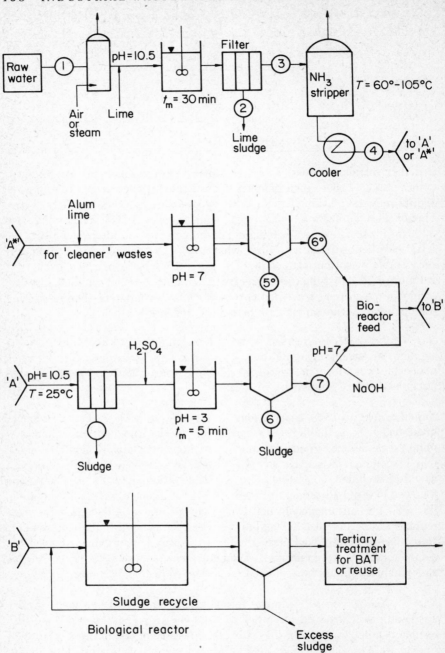

Fig. 2.10 Flow diagrams for pretreatment and biological treatment of fixed-bed coal conversion wastewaters (after Neufeld *et al.*, 1981c)

Step 4. Fixed ammonia stripping. Ammonia was steam or air stripped in the laboratory in electrically heated glass-lined batch reactors at temperatures of 60°–105°C. The wastewater was held in the stripper until the total ammonia in solution was reduced to the desired level, and then cooled in large containers. The pH of wastewater leaving the stripper was approximately within one pH unit of that entering the stripper. Although batch ammonia stripping is not used in commercial operations, the chemical evaluation data obtained in the laboratory should be scaleable; data on kinetic rate and economics would not be applicable.

If the wastewater was considered dirty, i.e., having a relatively low fraction of phenol carbon to total organic carbon in solution, the wastewater treatment step outlined by flowpath A would be followed. The objective of flowpath A is to maximize organic sludge production by means of judicious pH adjustment, coagulation and flocculation.

Flowpath A* was utilized with relatively cleaner wastewaters. The major difference between the two flowpaths is that pH depression below pH 7 was not accomplished in flowpath A*, thus less chemicals were used in the pretreatment train.

For dirtier condensates, the wastewater after ammonia stripping was filtered to retain a black silty sludge. Laboratory analysis of this sludge indicated that it contained an average of 70% to 75% volatile solids (Sample No. 1). Thus, most of the sludge removed at this step was organic in nature. For this dirtier wastewater, the pH was depressed to about 3.0 using sulphuric acid. At step 7, the wastewater was allowed to settle before filtration. The sludge developed was found to settle very rapidly and exhibit a composition of approximately 50% volatile material. Sodium hydroxide was then added to the wastewater with appropriate bio–trace nutrients so that final pH was approximately 7.0. This wastewater was then considered for feed to the biological reactors.

For cleaner wastewaters, the flow path outlined on A* was utilized. This step takes the cooled water from the ammonia stripper system, and goes through a standard coagulation flocculation scheme using lime and alum. After removal of the resultant sludge, appropriate nutrients were added and this waste too was considered for feed to the biological reactors.

Biological reactor phase

In this laboratory phase, pretreated biofeed was diluted as desired with tap water, and fed on a continuous basis to completely mixed activated sludge reactors with internal clarifiers. The bioreactors were held at sludge ages of 10–20 days, with *hydraulic retention times* of 1.0 day. Hydraulic times of less than one day may be achieved using this configuration. However, this approach was limited by the availability of feed to the overall experimental scheme, and ability to transfer oxygen when using conventional 'fish tank' air stones in laboratory reactors. Dilutions of up to 60% fixed-bed wastewater in tap water were successfully treated in a stable fashion using this technique without solvent extraction pretreatment. Toxic inhibition limitations, at 60% dilution, were not noticed. Technological limitations to using higher concentrations appear to be in the areas of oxygen transfer using laboratory systems, and sludge deflocculation. Sludge deflocculation is defined as the presence of pinpoint floc throughout the effluent which may or may not be accompanied by a sharp point of demarcation for settled sludge. Sludge bulking, on the other hand, exhibits a sharp demarcation between supernatant and the settled sludge phase, with supernatant being relatively clear but with a relatively poor degree of sludge settling. Sludge bulking indicates a high sludge volume index value, whereas sludge deflocculation may have either a high or a low sludge volume index.

Figure 2.11 plots TOC and phenol concentration at various points within the pretreatment train using the aged (worst case) coal conversion wastewater (sample No. 1). Comparison of these plots shows that almost 50% of the soluble TOC present in the influent wastewater was removed by the overall precipitation pretreatment step while only small levels of phenol were similarly removed. Solvent extraction, when used in the pretreatment step, removes mainly monohydric phenolics and does little for the non-biodegradable organic fraction.

In contrast to the above results obtained using dirtier gasifier wastewaters, Table 2.14 lists a characterization of sample No. 3 wastewater. With the cleaner wastewater, using flowpath A*, the overall TOC reduction was about 27%, with overall phenol reduction approximately 23% across the pretreatment train. Thus, the overall phenol/TOC ratio went from 70% in raw feed to 74% in the raw biofeed. The biokinetic data discussed in the next section relate to wastewater treated with analysis as outlined in Table 2.14.

2.6.3 Design of biological reactors

In order to function in a coal conversion facility wastewater treatment plant, a biological reactor must provide:

- stable operations;
- minimum hydraulic detention times, with corresponding minimum capital costs.

A properly designed pretreatment train may do much toward developing stable biological operations. However, many chemical–biochemical interactions that can occur within the biological plant may cause temporary toxic inhibition leading to bioreactor failure. These interactions include those of phenolics, ammonia compounds, thiocyanate compounds and other trace organics. Process design should aim to remove most, if not all, of these constituents in a one-stage or possibly a two-stage biological reactor configuration. As an example of toxic inhibition and the complex biochemistry that may exist in coal conversion waste treatment facilities, Neufeld *et al.* (1981a)

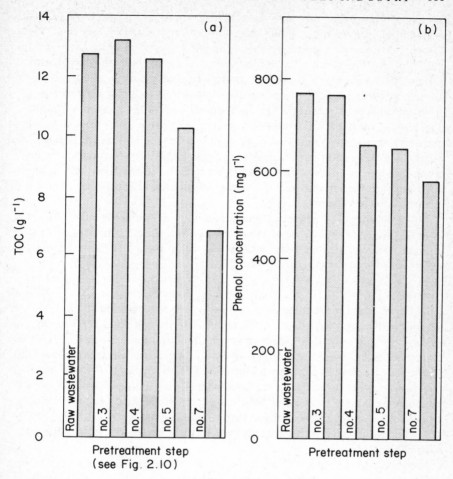

Fig. 2.11 (a) TOC levels and (b) phenol levels at various points in the pretreatment train (after Neufeld *et al.*, 1981c)
Point 3 After free ammonia stripping
Point 4 After fixed leg stripping (total TOC)
Point 5 After fixed leg stripping (after filtration)
Point 7 Bioreactor feed

quantified the biokinetics of inhibition of phenol degradation by the presence of trace levels of thiocyanate. An overall expression describing temperature effects and thiocyanate effects on phenol biokinetics has been developed showing profound effects when low levels of effluent phenol are desired.

Neufeld *et al.* (1981b) have developed data showing biokinetic degradation rates of thiocyanates that may be found in coal conversion and hot gas desulphurization blowdowns. Their work shows that thiocyanate biokinetics follows a classic substrate inhibition model whereby the rate of thiocyanate degradation is rapid until thiocyanate concentrations are in excess of about 470 mg l^{-1}, and then decreases with further increase in thiocyanate. They caution that difficulties may be expected in design for thiocyanate-laden wastewaters, particularly those wastewaters originating from sulphur removal processes where blowdown streams are often added to biological reactors.

Greenfield and Neufeld (1981) and Neufeld *et al.* (1984) present data on the interactions of non-biodegradable organics that have been shown to pass through a first-stage phenolic biological reactor on the biokinetics of nitrification. This work quantified the effect of simple methylated phenolics, ethylpyridine, thiocyanate, phenol and other substances on nitrification and incorporated this information into a series of rational engineering design and biological operational relationships.

Although phenol is inherently biodegradable, the interactions that may exist in coal conversion wastewaters, if not allowed for, have the potential to cause ultimate failure of what otherwise would be a highly efficient system for treatment of these residuals.

2.6.4 Pilot-plant bioreactor data — process stability and performance

As outlined above, very little research has been conducted to date on the biological treatability of coal conversion residuals, and much of what has been done was with limited quantities of coal conversion wastewaters or with synthetic wastewaters. Data by Luthy *et al.* (1979), Sack (1979) and Neufeld *et al.* (1978, 1981a) have shown that such phenol-containing wastewaters behave in somewhat similar ways and follow expected theoretical biokinetic trends. Values for constants of these 'theoretical kinetic equations' deviate considerably from one wastewater to another; however, most reported values fall within reasonable ranges.

Methods of evaluating treatability have been developed since the mid-1970s when coal conversion residuals were considered to contain such high levels of toxic organics as to preclude biological treatment. The state of the art today is that not only has it been demonstrated that coal conversion residuals may be biologically treated, but a rational and theoretical methodology has been developed for generic application. Questions still remain, however, with regard to material balance for heavy metal and trace organics necessary for water reuse applications and a rational basis for environmental regulations.

As an example of process stability which may be obtained, Figure 2.12 shows cumulative distribution plots of soluble total organic carbon (TOC) for two similar biological reactors treating 40% and 60% diluted fixed-bed coal conversion wastewaters by means of the pretreatment steps outlined above. As shown by the relatively shallow slope and parallel lines, the effluent distributions are quite similar in terms of small but similar standard deviation, and differ only in absolute concentrations. The data similarly show that a residual

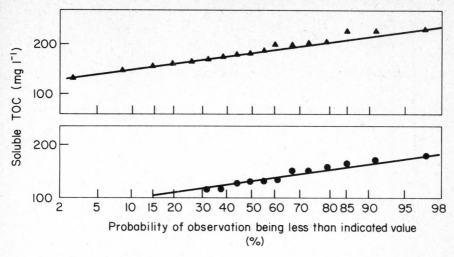

Fig. 2.12 Soluble TOC: log probability distributions for R3 (●) and R4 (▲) effluents (after Neufeld *et al.*, 1981d)

TOC exists in solution which appears not to be biodegradable under any circumstances. Indeed, we have found that approximately 10–15% of the METC soluble total organic carbon entering a biological reactor, even under the most favourable pretreatment conditions, is non-biodegradable and exits the reactor. Although virtually complete removal of all phenolics and BOD may be achieved, only 85–90% removal of soluble TOC was accomplished with METC wastewaters. Other fixed-bed coal gasification wastewaters may have higher or lower soluble nondegradable fractions. It is significant that the fraction of non-biodegradable species should be reported on the basis of fraction of influent rather than concentration. In experiments reported by Neufeld *et al.* (1978) and Johnson *et al.* (1977), various wastewaters from an assortment of Synthane coal gasification runs were utilized in both pretreatment and biological treatment. Analysis of the resulting data has shown that a single numerical concentration value to represent a non-biodegradable fraction is meaningless, and use of a percentage of influent concentration results in data that can be better correlated. This, however, requires more complex laboratory characterization than is commonly done.

Research with coal conversion residuals has shown that biokinetic evaluations often conform to classical Michaelis–Menten biokinetic relationships. Table 2.16 summarizes biokinetic relationships obtained to date for coal liquefaction and coal gasification wastewaters as determined in the laboratory. These kinetic relationships are given as:

$$q = \frac{Q_{max} S_e}{K_m + S_e} \quad \text{or} \quad q = K S_e$$

Table 2.16 Summary of biokinetic coefficients

MICHAELIS–MENTEN MODEL: $q = Q_{max} S_e/(K_m + S_e)$

Wastewater process	Substrate basis	Q_{max} (kg substrate per kg VSS per day)	K_m (mg substrate per litre)	Non-biodegradable fraction of influent (%)
Synthane[a]	BOD$_5$	0.46	5	15
	TOC	0.49	100	16
	COD	0.96	120	
	Phenol	0.63	30	
Synthoil liquefaction	BOD$_5$	0.39	14	10
	TOC	0.14	3	11
	COD	0.43	10	
METC fixed bed[b]	BOD$_5$	3.0	28.8	8–15

FIRST-ORDER MODEL: $q = KS_e$

Wastewater process	Substrate basis	K (kg substrate removed per kg VSS per day)
METC fixed bed[c]	BOD$_5$	0.35–0.6
GFETC fixed bed[d]	BOD$_5$	0.14–0.32
	COD	0.21–0.48
Simulated composite wastewater[e]	BOD$_5$	0.18–0.71
	COD	0.29–1.13

[a] Drummond et al. (1980).
[b] Neufeld et al. (1981a).
[c] Sack (1979).
[d] Luthy (1980).
[e] Singer (1979).

where K is the first-order rate constant
K_m is a constant (mass/volume)
q is the specific substrate utilization rate per unit biomass (mass/mass–time)
Q_{max} is a constant (mass/mass–time)
S_e is the effluent substrate concentration (mass/volume)

Biokinetic yield coefficients were found to follow the relationship

$$dX/dt = a(dS/dt) - bX$$

where a is the yield coefficient (mass/mass)
b is the decay coefficient (time^{-1})
S is the biodegradable substrate concentration (mass/volume)
t is time mass/volume
X is the biomass concentration (mass/volume).

Table 2.17 summarizes yield coefficients and decay coefficients found to date in the literature.

Biomass respiration relationships indicate the amount of oxygen required to stabilize organics. The relationship often used to calculate oxygen requirement is

$$dO_2/dt = a'(dS/dt) + b'(X)$$

Table 2.17 Summary of yield coefficient and decay coefficient data: $dX/dt = a(dS/dt) - b(X)$

Wastewater source/substrate	Basis	a (kg VSS per kg substrate)	b (day^{-1})
Synthane[a]	BOD$_5$	0.367	0.033
Synthane[b]	BOD$_5$	0.53	0.02
	TOC	1.1	0.04
	COD	0.47	0.05
Synthoil liquefaction[b]	BOD$_5$	0.61	0.02
	TOC	1.2	0.004
	COD	0.36	0.004
METC fixed bed[c]	BOD$_5$	0.54	0.09
	TOC	1.40	0.14
METC fixed bed[d]	BOD$_5$	0.24	0.062
GFETC fixed bed[e]	COD	0.29	0.038
Simulated composite wastewater[f]	BOD$_5$	0.27 (SS basis)	0.005
	TOC	0.52 (SS basis)	0.005
	COD	0.18 (SS basis)	0.005

[a] Neufeld et al. (1978)
[b] Drummond et al. (1980).
[c] Sack (1979).
[d] Neufeld et al. (1981a).
[e] Luthy (1980).
[f] Singer (1979).

Table 2.18 Summary of oxygen utilization coefficient data: $dO_2/dt = a'(dS/dt) + b'(X)$

Wastewater source/substrate	Basis	a' (kg O_2 per kg substrate)	b' (kg O_2 per kg VSS)
Synthane[a]	BOD_5	0.747	0.110
	TOC	1.680	0.090
	COD	0.562	0.093
Synthane[b]	BOD_5	0.86	0.03
	TOC	1.9	0.05
	COD	0.63	0.04
Synthoil liquefaction[b]	BOD_5	1.2	0.02
	TOC	2.1	0.01
	COD	0.60	0.01
METC fixed bed[c]	BOD_5	0.30	0.08
	TOC	0.67	0.08
METC fixed bed[d]	BOD_5	0.80	0.038
GFETC fixed bed[e]	BOD_5	1.0	0.05
	COD	0.77	0.01

[a] Neufeld et al. (1978).
[b] Drummond et al. (1980).
[c] Sack (1979).
[d] Neufeld et al. (1981a).
[e] Luthy (1980).

where a' is the oxygen respiration rate coefficient ($m\,O_2/(m$ biomass $- t$))
b' is the oxygen use during decay ($m\,O_2/(m$ biomass $- t$)).

Published values for a' and b' of coal conversion effluents are given in Table 2.18.

2.7 Wastewater treatment costs for coal conversion facilities

2.7.1 Basis of cost assessment

In June 1979 the US DOE, Office of Environmental Compliance and Overview, held a workshop on processing needs and methodology for wastewaters from the conversion of coal, oil shale and biomass to synfuels (DOE, 1980). As part of the workshop output, relative costs associated with processing of synfuels wastewater were explored. Although many factors affect the selection and costs of various processes, it should be noted that for the most part, as with municipal wastewater treatment facilities, treatment costs are more a function of volume of water to be treated rather than the concentration of pollutants in the water stream. Thus, costs are usually better described in terms of volume treated rather than plant capacity or nature of wastewater produced.

With these limitations in mind, an estimate of costs associated with various processes is given below. It should be noted that the costs indicated are only rough estimates for representative wastewater treatment systems. More recent cost estimates of specific unit operations are given in EPA (1983). The real costs of any system are dependent upon the type of gasifier employed (phenolic producer or nonphenolic producer), environmental regulations in force at the time, potential for recovery of phenolics and ammonia, fraction of organic substances that may be attributable to phenolics, availability of oxygen for advanced treatment, availability and costs of land, and use of discharged water (i.e. is water reused in the process, or is water suitable and acceptable for discharge to the environment?).

2.7.2 Gravity separation techniques
Gravity separation techniques include settling ponds, API separators, centrifuges, etc. Costs: $0.40–$0.80 per m^3 water treated (DOE, 1980).

2.7.3 Steam stripping
Steam stripping can remove the major portion of the NH_3, CO_2, H_2S and HCN dissolved in water. It would be necessary to adjust the pH of solution prior to ammonia stripping to remove fixed ammonia as required to meet BAT standards for synthetic fuels facilities. Byproduct ammonia recovery will be attractive for virtually all coal conversion facilities. Costs: about $38 per m^3 (not including credits for byproduct ammonia recovery) (DOE, 1980).

2.7.4 Solvent extraction
Research at major US universities, coupled with data reported in the literature, shows that solvents such as methylisobutylketone (MIBK) and diisopropyl ether (DIPE) are useful in extraction of phenols, polyhydric aromatics and other trace substances. Solvent extraction permits realization of the potential for byproduct recovery, provided that the pH range is acceptable. Lowering of the pH to below 8.0 may be necessary for the removal of many organic acids including trihydric phenols. In most cases, solvent extraction must be accompanied by a solvent recovery step and limitations on solvent losses in the aqueous phase place a technical and economic limit on choices and characteristics of acceptable solvents. Costs: about $19.00 per m^3 (DOE, 1980).

2.7.5 Biological treatment
Biological treatment has a long history of successful application to both municipal and industrial wastewaters. By definition, this technique should remove essentially all biodegradable materials and also a substantial fraction of adsorbable substances. Research outlined above shows that biological treatment

of coal gasification wastewaters can result in approximately 85–95% removal of total soluble organic carbon and total soluble COD.

Costs for municipal wastewater treatment are relatively low. However, the costs for coal refineries, because of the relatively high BOD of coal conversion wastewaters, are estimated (DOE, 1980) to be $38–$76 per m^3 for influent BOD equal to 10 000 mg l^{-1}. The costs of biological treatment are dependent upon availability of land and energy, i.e., low levels of energy (lagoons, etc.) may be utilized if sufficient land space is available, while high-energy systems such as those using pure oxygen may be required if the costs of land are high.

2.7.6 Carbon adsorption

Activated carbon adsorption of treated biological effluents may be required by future legislative constraints. Adsorption by activated carbon can result in major reductions in colour and soluble TOC. Activated carbon costs are directly related to wastewater organic levels, which are governed by the efficiency of organic removal in the pretreatment system and biological system.

Based on 2000 mg l^{-1} COD, and 6% carbon loss during regeneration, DOE (1980) estimates the costs of activated carbon to be in the range of $23.00–$45.00 per m^3 treated.

2.7.7 Wastewater treatment train

It is estimated that representative wastewater treatment train costs for a coal conversion facility may be in the neighbourhood of $26.00–$114.00 per m^3 if the wastewater treatment train consists of pretreatment with ammonia recovery, solvent extraction and biological oxidation using air. Should the treatment step require further oxidation and advanced treatment such as ozonation and activated carbon, it is possible that the costs for wastewater treatment may approach $190.00 per m^3 wastewater treated (1980 US dollars). Figure 2.13 summarizes representative wastewater treatment train costs for coal conversion processes.

2.8 Solid waste generation, treatment and disposal

2.8.1 Sources of coal conversion residuals

The solid wastes generated by coal conversion processes that exert the greatest influence on residuals management efforts are (Bern *et al.*, 1980; Neufeld, 1981e):

- Coal preparation plant residuals
- Gasifier ash or slag
- Wastewater treatment plant sludges
- Auxiliary flue gas desulphurization sludges

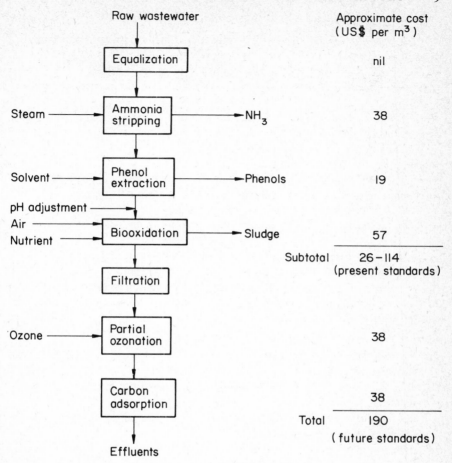

Fig. 2.13 Representative wastewater treatment train for coal conversion effluents (after DOE, 1980)

These residuals, which have the potential to leach trace organics and heavy metals into the groundwater, may be potentially toxic and may exhibit potentially carcinogenic properties.

Public opposition to a massive synthetic fuels programme is anticipated to be based upon potential or perceived adverse effects on health. Potentially toxic heavy metals are present in all feed coals, at least in trace amounts. The fate of trace substances undergoing the coal conversion process and the probability of their migration from final disposal sites into the open environment are a primary concern when considering environmentally acceptable disposal options.

Specific coal conversion solid waste streams of concern are:

- *Coal preparation plant area* — coal refuse, coal dust and wastewater from the tailings pond.
- *Coal gasification plant area* — residual ash and quench-water.
- *Steam and power plant generation area* — residual ash and flue gas desulphurization sludges.
- *Raw water treatment area* — sludge from solids in the raw water source.
- *Wastewater treatment plant area* — lime sludge, organic sludges, waste biological sludges and oil and tar residuals.
- *Tar separation area* — sludges.
- *Phenol removal area* — filter backwash and sludges containing phenolics.
- *Sulphur removal area* — elemental or product sulphur if not saleable.
- *Tailgas treatment area* — residual sulphur sludge.
- *Spent catalysts* — from shift and product upgrading areas.

Land disposal for solid residuals is economically imperative in an overall management scheme of synfuels solid wastes. Such land disposal, however, must be done in an environmentally acceptable manner. Bern *et al.* (1980) outline management alternatives that are available to owners of commercial synthetic fuels facilities. Their report includes presentation of existing data furnishing:

(a) an understanding of available and feasible alternatives at large-scale coal conversion facilities;
(b) an analysis of the various innovative concepts for management of such residuals;
(c) development plans of a waste management system in the framework of an idealized commercial-scale plant;
(d) delineation of areas for which further study and research are most critical for developing management options for solid wastes.

2.8.2 Gasifier ash or slag

Neufeld *et al.* (1981e) outline chemical and biological properties of coal conversion ash residuals derived from US Government-sponsored large-scale coal conversion facilities and direct liquefaction facilities. This characterization of such solid waste residuals included development of natural particle size distributions and heavy metal analysis of leachates from each size of fraction. Results show that smaller fractions yield much greater quantities of heavy metal in derived leachates. Leaching procedures were conducted in accordance with the US EPA's 'acetic acid extraction procedure' (EPA–'EP') which complies with requirements for RCRA 'hazardous waste' in US Federal regulations. In addition, solid waste samples were leached in accordance with the ASTM (American Society for Testing and Materials) Method A distilled deionized water leaching procedure. All leaching procedures are summarized by Neufeld et al. (1981e). In no case did resulting leachates contain more than one hundred times the heavy metals concentration of primary drinking water—a value above which wastes are deemed by the RCRA to be 'hazardous'. The leaching techniques employed, however, did yield significant quantities of iron in the resulting leachates. This iron is thought to be derived from large quantities of tramp iron found in slagging coal gasification solid residuals when using Western USA lignite coal as feedstock.

Of the samples studied to date, in no case did leachates from coal conversion solid waste ash residuals give positive results in the *Salmonella*–Mammalian Microsome (Ames) Assay. A positive result from Ames testing indicates potential for mutagenicity in tested material. Evidence, however, of *Daphnia* toxicity was observed in some derived leachates.

Ash residuals are expected to be by far the largest volume of solid wastes generated at a coal conversion process. While the levels of heavy metals leached from coal conversion residuals and their potential for mutagenicity is minimal, geotechnical considerations and questions still exist regarding their fine grain structure, compactability and process indices. Evaluation of such indices is essential for development of a rational geotechnical approach to engineered landfill.

2.8.3 Wastewater treatment plant sludges

As outlined on Figure 2.10, a variety of sludges are expected to be generated during the course of wastewater treatment operations. These sludges include:

- Lime sludges resulting from pH adjustment prior to ammonia stripping.
- Organic sludges resulting from filtration prior to biological oxidation.
- Biological sludges resulting from the growth of excess organisms during biological treatment.

If a filtration step is not included between ammonia stripping and biological oxidation, the insoluble organic material that would otherwise be precipitated will be removed in the excess biosludge. The disadvantage of this technique is that such trace organics may prove inhibitory to the biological reaction phase. Additional sludges may be generated by tertiary treatment of gasification wastewaters. Such sludges may include rapid sand filter backwashes and cooling tower sludges and blowdowns where treated or partially treated gasification wastewaters are used as cooling tower makeup.

Research conducted at the University of Pittsburgh utilizing coal gasification wastewaters from the US DOE–Morgantown Energy Technology Center has produced sludges when waters were treated in accordance with the pretreatment flow diagram shown in Fig. 2.10. Figure 2.14 shows the relative quantities and compositions of sludges produced when treating a sample of METC coal gasification wastewater derived from 'Sample No. 3' (see Table 2.14) utilizing

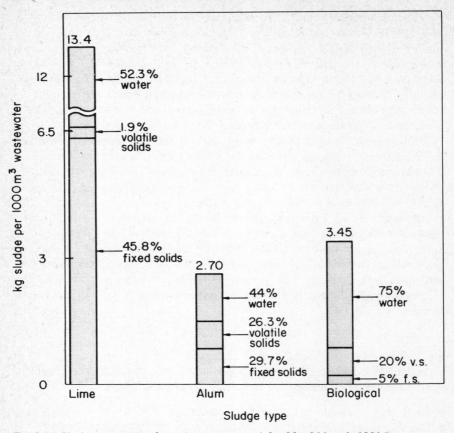

Fig. 2.14 Sludge generation for various processes (after Neufeld *et al.*, 1981d)

pretreatment step A* and subsequent biological treatment. The quantities of sludge produced from the wastewater treatment step are presented on a normalized basis of mass of sludge per mass of wastewater treated. As may be seen from Fig. 2.14, approximately 12 tonnes of wet (52.3% water) lime sludge are produced per 1000 m^3 of wastewater treated. This extremely high level is due partially to the high level of alkalinity exhibited by METC gasification wastewaters. Figure 2.14 also shows that about 2.6 tonnes of wet sludge (at 44% moisture) may be expected to be produced per 1000 m^3 of wastewater treated. The organic content of the alum sludge is considerable, the fraction of total solids considered as organic being approximately 47% [26.3/(26.3 + 29.7)]. The resulting alum sludge is blackish-brown in colour and has a phenolic-type odour.

In addition, it is anticipated that about 3.5 tonnes of waste biological sludge will be produced per 1000 m^3 of undiluted wastewater treated. A value of 25% total solids is a reasonable estimate for vacuum-filtered waste biological sludges; this value was used in computation of total disposable biosludges shown in Fig. 2.14.

Table 2.19 lists laboratory-derived data for TOC, phenolics and heavy metals that appeared in leachates resulting from the use of the EPA–'EP' and ASTM–A leaching procedures on both lime sludges and organic alum sludges derived from wastewater treatment. As can be seen from this table, considerable levels of calcium appear in all derived leachates. This may be expected with the high levels of lime used for ammonia stripping. Substantial conductivity results from the ASTM–A procedures (distilled deionized water technique). In view of the calcium level in leachates, such conductivity is to be expected. It is interesting to note, however, the high level of phenol in leachates as measured by the 4-amino antipyrine technique. The distilled deionized water technique (ASTM–A) appeared to extract greater quantities of phenolics than did the US EPA–'EP' procedure. The presence of such phenolics has a strong bearing on environmental acceptability since there is a possibility of contamination of groundwater.

Such data must be considered as preliminary at the present time; research is under way at the University of Pittsburgh and US DOE to evaluate potential effects on health by means of toxicity testing, mutagenicity testing and advanced organic and trace metal analysis on additional sludges derived from the pretreatment train and its modifications.

2.8.4 Control options for solid waste disposal and leachate treatment

As outlined by Neufeld *et al.* (1981e), management of solid ash–slag waste residuals from the gasification or coal liquefaction area forms only one part of the overall solid waste management programme for the emerging synthetic fuels industry. However, it is this portion which has been emphasized in all solid waste research conducted to date.

Pretreatment of some process residuals may be necessary prior to disposal. The sludge generated by raw water processing, which contains mainly grit, settled solids and filter sludges is generally composed of inert materials which may be landfilled without pretreatment. Boiler blowdown and cooling tower blowdown containing high concentrations of dissolved salts, chromium or organic-based algicides and other trace metals are likely to require some pretreatment prior to disposal.

The liquid, semisolid and solid residuals resulting from coal gasification operations normally require pretreatment prior to disposal. The exception may be fly ash, bottom ash and slags. Conditioning of residuals may take the form of chemical or biological treatment to remove phenolics and other trace organics, followed by encapsulation of salts which would otherwise be leachable.

Surface water runoff from coal mining and cleaning plant refuse needs chemical neutralization prior to discharge to surface waters. Settling ponds may be required for removal of coal refuse fines and solids from wash-water.

Table 2.19 Wastewater sludge leachates: typical and preliminary data

Sludge type	Leaching test	Sludge leachates				Heavy metals in leachates (mg l^{-1})					
		Final pH	TOC (mg l^{-1})	Phenol (mg l^{-1})	Conductivity (μmhos)	Fe	Zn	Cd	Cu	Mg	Ca
Lime sludge	EPA–'EP'	7.29	2630	155	6400	0.8	0.24	0.1	0	26	1620
	ASTM–A	11.84	1590	720	4800	1.4	0	0	0	0.13	640
Organic 'alum' sludge	EPA–'EP'	6.92	2800	455	7100	0.8	0.44	0	0.3	26	1800
	ASTM–A	7.75	1960	1550	5300	7.0	0	0	0	176	340

Further treatment such as filtration of wash-water may be necessary to meet local water quality effluent standards. It is possible that solid waste streams may be physically or chemically altered to promote stabilization with hardened sludges. Such treatment may be achieved by the addition of cementitious or 'fixing' materials to semisolid sludges and slurries. If leaching characteristics and permeability of such residuals are reduced, it is possible that the rate of release of toxic and hazardous materials may be drastically decreased to acceptable concentration levels, allowing a practical approach to disposal of solid residuals.

In summary, from a pragmatic viewpoint, the ranking of disposal alternatives would be as follows:

(1) Direct landfill with codisposal of all residual streams on an on-site disposal location.
(2) Direct land disposal of all residual streams on-site, with modification of hydrogeological factors by means of engineering design or linings to landfill so that leachate may be isolated and treated.
(3) Off-site disposal with modification as in (2).
(4) Pretreatment of individual waste streams as necessary to modify physical and chemical behaviour prior to disposal.
(5) Chemical treatment that modifies or chemically changes or destroys the waste, thus rendering it pollution-free prior to land disposal.
(6) Destruction of some waste streams in the gasification scheme by high-temperature incineration.

The physical state of residuals, such as the water content of a sludge or particle size distribution of a residue, limits the use of some disposal sites because of the possible instability of wastes when placed in certain settings (geotechnical considerations). Adverse engineering properties can lead to slide conditions, and failure of berm and embankments may result in escape of entrained material. The dominating factor, however, that governs the disposability of residues is the large volumes involved. The chemical and physical nature of the waste, pollution potential and degree of handling are almost secondary.

2.9 Summary

An overview of coal conversion processes is presented in the detail required by the modern environmental engineering designer. Data collected to date illustrating compositions and normalized pollution production parameters for synfuels wastewaters are shown. Wastewater treatment flow diagrams that are applicable to a variety of synfuels effluent streams are developed with data relating to sludge production parameters, conditions of wastewater treatment, process kinetics and process economics.

It is possible to achieve environmental acceptability of synthetic fuels facilities only when potential environmental discharges are properly recognized, evaluated and taken into account in design. Environmental acceptability for any facility costs money; however, failure to account for environmental acceptability for a synfuels facility may cost society more than just money. The challenge of a clean synthetic fuels plant must be met equally by owners, investors, designers, governments and neighbours in a rational pragmatic fashion if the industry is to evolve from its current embryonic state.

References

APHA (1980) *Standard Methods for the Examination of Water and Wastewater*, 15th edn, APHA, AWWA, WPCF, Washington, DC.

Bern, J., Neufeld, R.D. and Shapiro, M.A. (1980) *Solid Waste Management of Coal Conversion Residuals from a Commercial Sized Facility: Environmental Engineering Aspects*, Report DOE/ET/20023−5, US Department of Energy, Pittsburgh Energy Technology Center.

Bertrand, R.R., McGee, E.M., Janes, K.T. and Rhodes, W.J. (1974) Trip report (US EPA RTP/IERL)—Four Commercial Gasification Plants—November 7−18, unpublished.

Bromel, M.C. and Fleeker, J.R. (1976) *Biotreating and Chemistry of Wastewaters from the South African Coal, Oil and Gas Corporation (Sasol) Coal Gasification Plant*, internal report, Department of Bacteriology, North Dakota State University, Fargo, North Dakota, as referenced by Singer *et al.* (1978).

Cleland, J.G. and Kingsbury, G.L. (1977) *Multimedia Environmental Goals for Environmental Assessment*, Report EPA-600/7-77-136a, US Environmental Protection Agency, Washington, DC.

Cooke, R. and Graham, P.W. (1965) Biological purification of the effluent from a Lurgi plant gasifying bituminous coals, *Intl J. Air Water Pollut. (England)*, 9(3), 97.

DOE (1980) *Processing Needs and Methodology for Wastewaters from the Conversion of Coal, Oil Shale and Biomass to Synfuels*, Report DOE/EV-0081, Office of Environmental Compliance and Overview, Washington, DC.

Drummond, C.J., Johnson, J.E., Neufeld, R.D. and Haynes, W.P. (1980) Biochemical oxidation of coal conversion wastewaters, in *Water — 1979*, AIChE Symp. Series, Bennett, G.F. (Ed.) Vol. 76, 209−214.

EPA (1980) *Development Document for Proposed Effluent Limitation Guidelines, New Source Performance Standards, and Pretreatment Standards for the Iron and Steel Manufacturing Point Source Category*, Report 440/1-80/024, US Environmental Protection Agency,

EPA (1983) *Control Technology Appendices for Pollution Control Technology Manuals*, Report EPA 600/8-83-009, US Environmental Protection Agency, Washington, DC.

Ghassemi, M., Crawford, K. and Quinlivan, S. (1978) *Environmental Assessment Data Base for High BTU Gasification Technology*, Vol. 2, Appendices A, B and C, Report EPA-600/7-78-186b, US Environmental Protection Agency, Washington, DC.

Ghassemi, M., Crawford, K. and Quinlivan, S. (1979) *Environmental Assessment Report: Lurgi Coal Gasification Systems for SNG*, Report EPA-600/7-79-120, US Environmental Protection Agency, Washington, DC.

Greenfield, J. and Neufeld, R.D. (1981) Quantification of the influence of steel industry trace organic substances on biological nitrification, Proc. 36th Purdue Ind. Waste Conf., Ann Arbor Sci. Publ., Ann Arbor, Michigan, 772–783.

Heredy, L.A. and Wender, I. (1980) Model structure for a bituminous coal in *American Chemical Society Division of Fuel Chemistry, preprints of papers presented at San Francisco, California*, **23**(4), 38–45. Available from American Chemical Society, Washington, DC.

Hossain, S.M., Cilione, P.F., Cherry, A.B. and Wasylenko, W.J. (1979) *Applicability of Coke Plant Control Technologies to Coal Conversion*, Report EPA-600/779-184, available from the Industrial Environmental Research Laboratory, US Environmental Protection Agency, Research Triangle Park, North Carolina.

Hovious, J.C. (1972) *Anaerobic Treatment of Synthetic Organic Wastes*, Water Pollution Control Research Series, Report 12020DIS01/72, US Environmental Protection Agency, Washington, DC.

Iglar, A.F. (1980) *Study of the Treatability of Wastewater from a Coal Gasification Plant*, Final Report DOE/ET/00234/31, US Department of Energy, Pittsburgh Energy Technology Center.

Janes, K.T. (1980) *Draft Pollution Control Guidance Document Indirect Liquefaction*, Vol. II, *Data Base* and Vol. III, *Appendices*, US Environmental Protection Agency, Industrial Environmental Research Laboratory, Research Triangle Park, North Carolina. Superseded by EPA (1983) *Pollution Control Technical Manual for Lurgi Based Indirect Liquefaction*, EPA-600/8-83-006, and *Pollution Control Technical Manual for Kopper–Totzek Based Indirect Liquefaction*, EPA-600/8-83-008, US Environmental Protection Agency, Washington, DC.

Johnson, C.A. *et al.* (1975) Present status of the H–Coal process in clean fuel, *Gas Symposium II*, Institute of Gas Technology, Illinois Institute of Technology Center, Chicago, Illinois.

Johnson, G., Neufeld, R.D., Drummond, C., Strakey, J., Haynes, W. and Mack, J. (1977) *Treatability Studies of Condensate Water from Synthane Coal Gasification*, Report DOE-PERC RI 77/13, US Department of Energy, Pittsburgh Energy Technology Center, USA.

Luthy, R.G., Sekel, D.J. and Tallon, J.T. (1979) *Biological Treatment of Grand Forks Energy Technology Center Slagging Fixed Bed Coal Gasification Process Wastewater*, Report FE-2496-42, US Department of Energy, Washington, DC.

Luthy, R.G. and Tallon, J. (1978) *Experimental Analysis of Biological Oxidation Characteristics of Hygas Coal Gasification Wastewaters*, Report FE-2496-43, US Department of Energy, Washington, DC.

Luthy, R.G. (1980) Biological treatment of synthetic fuel wastewater, *Proc. ASCE*, **EE3**, 609.

McKenna, E.J. and Heath, R.D. (1976) *Biodegradation of Polynuclear Aromatic Hydrocarbon Pollutants by Soil and Water Micro-organisms*, Research Report 113, Illinois Water Resources Center (available from US NTIS, PB25364).

Neufeld, R.D., Drummond, C. and Johnson, G. (1978) Biokinetics of activated sludge treatment of synthane fluidized bed gasification wastewaters, *Am. Chem. Soc., Preprints — Fuels Division — 175th National Meeting*, **23**(2), 175–186.

Neufeld, R.D., Greenfield, J., Hill, A.J., Reider, C.B. and Adekoya, D.O. (1984) *Nitrification Inhibition Biokinetics*, Final Report to the US Environmental Protection Agency under contract No. CR807 527 (in press).

Neufeld, R.D., Mack, J.D. and Strakey, J.P. (1980) Anaerobic phenol biokinetics, *J. Water Pollut. Cont. Fed.*, **52**(9), 2367–2377.

Neufeld, R.D., Mattson, L. and Lubon, P. (1981a) Thiocyanate bio-oxidation kinetics, *J. Environ. Eng. Div. Proc. ASCE*, **107** (EE5), 1035–1049.

Neufeld, R.D., Mattson, L. and Lubon, P. (1981b) *Bio-oxidation of Thiocyanates Typical of Coal Conversion Effluents*, Report DOE/ET/4502-7, US Department of Energy, Washington, DC.

Neufeld, R.D., Moretti, C. and Ali, F. (1981c) Pretreatment and biological treatment of fixed bed coal gasification wastewaters, presented at the 36th Purdue Ind. Waste Conf., West Lafayette, Indiana.

Neufeld, R.D., Moretti, C., Ali, F. and Paladino, J. (1981d) *Pretreatment and Biological Digestion of METC Fixed Bed Gas Producer Wastewaters*, Final Report to US DOE No. DOE/ET/14372-7.

Neufeld, R.D. and Spinola, A.A. (1978) Ozonation of coal gasification plant wastewaters, *Environ. Sci. Technol.*, **12**(4), 470–472.

Neufeld, R.D. and Valiknac, T. (1979) Inhibition of phenol biodegradation by thiocyanate, *J. Water Pollut. Cont. Fed.*, **51**(9), 2283–2291.

Neufeld, R.D., Wallach, S.J., Erdogen, H. and Bern, J. (1981e) *Chemical and Biological Properties of Coal Conversion Solid Wastes*, Report DOE/ET/10061-1, US Department of Energy, Grand Forks Energy Technology Center, North Dakota.

Parker, C.L. and Dykstra, D.I. (1978) *Environmental Assessment Data Base for Coal Liquefaction Technology:* Vol. II *Synthoil, H–Coal and Exxon Donor Solvent Processes*, Report EPA-600/7-78-184b, US Environmental Protection Agency, Washington, DC.

Reap, E.J. *et al.* (1979) Wastewater characteristics and treatment technology for liquefaction of coal using H–Coal process, Proc. 32nd Purdue Ind. Waste Conf., Ann Arbor Science, Ann Arbor, Michigan.

Sack, W.A. (1979) *Biological Treatability of Gasifier Wastewater*, Report METC/CR-79/24, US Department of Energy, Morgantown Technology Center, Morgantown, West Virginia.

Sarna, K.R. and O'Leary, D.T. (1979) *Engineering Evaluation of Control Technology for the H–Coal and Exxon Donor Solvent Processes*, Report EPA-600/7-79-168, US Environmental Protection Agency, Washington, DC.

Schmidt, C.E., Sharkey, A.G. and Fridel, R.A. (1974) *Mass Spectrometric Analysis of Product Water from Coal Gasification*, TPR No. 86, US Bureau of Mines, Pittsburgh Energy Technology Center.

Singer, P.C. (1979) Evaluation of coal conversion wastewater treatability, in *Symp. Proc. Env. Aspects Fuel Conversion Technol.*, Ayer, F.A. and Jones, S. (Eds), EPA-600/7-79-217, US Environmental Protection Agency, Washington, DC, 457–478.

Singer, P.C., Lamb, III, J.C., Pfaender, F.K. and Goodman, R. (1979) *Treatability and Assessment of Coal Conversion Wastewaters: Phase I*, Report EPA-600/7-79-248, US Environmental Protection Agency, Industrial Environmental Research Laboratory, Research Triangle Park, North Carolina.

Singer, P.C., Pfaender, F.K., Chinchili, J., Maciorowski, A.F., Lamb, III, J.C. and Goodman, R. (1978) *Assessment of Coal Conversion Wastewaters: Characterization and Preliminary Biotreatability*, Report EPA-600/7-78-181, US Environmental Protection Agency, Industrial Environmental Research Laboratory, Office of Research and Development, Washington, DC.

Sinor, J.E. (1977) *Evaluation of Background Data Relating to New Source Performance Standards for Lurgi Gasification*, Report EPA-600/7-77-057, US Environmental Protection Agency, Industrial Environmental Research Laboratory, Research Triangle Park, North Carolina.

Stamoudis, V.C. and Luthy, R.G. (1980) *Biological Removal of Organic Constituents in Quench Waters from High BTU Coal Gasification Pilot Plants*, Argonne National Laboratories Report ANL/PAG-2, USA.

Stamoudis, V.C., Luthy, R.G. and Harrison, W. (1979) *Removal of Organic Constituents in a Coal Gasification Process Wastewater by Activated Sludge Treatment*, Argonne National Laboratories Report ANL/WR-79-1, USA.

Stearns-Rogers, Inc. (1980) *Fixed Bed Slagging Gasifier Pilot Plant, Gasifier Run #R.A.73 Report*, Report 19667, US Department of Energy, Grand Forks Energy Technology Center, North Dakota.

3 The treatment of wastes from the petrochemical industry

G E Chivers, *Centre for Extension Studies, Loughborough University of Technology*

3.1 The petrochemical industry

3.1.1 Introduction

The province of the petrochemical industry is the production of chemicals from petroleum feedstocks. However, the rapid growth of the industry since World War II, the wide range of processes carried out and the integration of oil refinery operations with downstream chemical processing makes strict definition difficult. Thus 'cracking' processes aimed at achieving higher-octane-number gasoline might be defined as refinery operations, while the same processes aimed at producing more chemical feedstocks might well be viewed as part of petrochemical processing.

In this review the petrochemical industry will be taken to exclude oil refining but include the production of basic organic chemicals from refinery products. Wastewater management for petroleum refineries is covered in Chapter 1 of this volume.

Because of the huge volumes of raw materials involved and the lower costs of transporting them by water, oil refineries and petrochemical works are most commonly located along the coasts of developed countries. Estuaries of major rivers are particularly favoured as they can provide sheltered deep-water harbours for tankers. Undoubtedly, a great deal of the coastal water pollution arising from the oil and petrochemical industry does not stem from landbased operations but from the tankers themselves. Unless the operations of tankers are closely controlled to minimize pollution, a great deal of money can be invested in treatment of refinery and petrochemical works wastewaters with little measurable improvement in environmental conditions. Similarly, handling and storage arrangements at the dockside must be designed to minimize pollution by spills and leaks.

Historically, there have been limited legal controls, if any, over the discharge of refinery and petrochemical plant wastewaters into coastal waters or the lower tidal reaches of estuaries (McCaul and Crossland, 1974). However, recent years have seen greatly increased concern about marine and coastal pollution and much stricter standards for wastewater discharges are being introduced, particularly with respect to immiscible oil and solvents and toxic constituents.

Despite the advantage of coastal location, many refineries and petrochemical works are located at inland sites adjacent to rivers, lakes and other surface waters. England and Wales alone have some fifty inland refineries discharging nearly 2 million m^3 of wastewater per day into rivers and canals (Department of Environment, 1975). Countries with large inland rivers used for bulk transportation have large numbers of major refineries and petrochemical works located inland. Thus, in the USA a huge range of such works line the banks of the lower Mississippi River, while the same is true for the Rhine and its tributaries in Western Europe. The legal standards for wastewater discharges into inland surface waters are usually much higher than for coastal water discharges, and wastewater treatment costs may be high.

3.1.2 Petrochemicals and petrochemical processes

From the early period of oil production for fuels, attempts were made to replace coal-derived chemicals with oil-derived chemicals. The first successful work on an industrial scale was carried out nearly 50 years ago in the USA by the Standard Oil Company of New Jersey which produced acetone from refinery gases (Stephens, 1970).

The petrochemical industry was mainly based in the USA before World War II, since most oil refineries were located there. Thermal cracking of oil to make gasoline produced refinery gases containing olefins which could be converted to useful chemicals. The USA was also favoured by access to natural gas which provided cheap energy and a source of ethylene.

After 1945 a policy of building refineries near the centres of consumption was adopted and this made possible a rapid growth in petrochemical production in Europe (about 20% per annum through the 1950s and 1960s). Outside the USA the principal feedstock for petrochemicals has been liquid rather than gas. Most commonly, petrochemical plants have been located adjacent to oil refineries or integral with them. Sometimes oil feedstocks are transferred from oil refineries to petrochemical plants at more distant locations. Typically both are located together on estuaries around Britain, and other European countries.

In this field the main processes by which the petroleum hydrocarbons are converted into raw materials for the chemical industry may be conveniently classified as follows:

(1) Removal of hydrogen, leading to unsaturated hydrocarbons (alkenes or aromatics): these unsaturated bonds are a source of strain and serve as points of attack for subsequent operations which yield useful end-products.
(2) Oxidation.
(3) Chlorination.

Removal of hydrogen to give alkenes

This process is by far the most important route by which oil feedstocks are converted into chemicals. Cracking at 650°–850°C produces ethylene, propylene, butenes and butadiene. To minimize their further reaction with each other they are diluted by introducing steam into the feedstock stream as mentioned earlier, and as a further step the products of cracking are rapidly cooled.

An example of this process is the conversion of normal butane into butadiene by dehydrogenation at about 65°C in the presence of a catalyst:

$$CH_3-CH_2-CH_2-CH_3 \xrightarrow{-H_2} CH_2=CH-CH_2-CH_3$$

$$n\text{-butane} \qquad \qquad \text{but-1-ene}$$

$$\downarrow -H_2$$

$$CH_2=CH-CH=CH_2$$

$$\text{buta-1,3-diene}$$

Ethylene is a product of major importance to the chemical industry, with some 16 million tonnes of ethylene plant capacity in Europe in 1980. This compound is converted to polythene, ethylene oxide, styrene, ethanol (ethyl alcohol), acetaldehyde and dichloroethane. Propylene is converted mainly to isopropanol, isobutanol, cumene, propylene oxide, polypropylene, glycerol and acrylonitrile. Under more severe cracking conditions (1000°–1600°C) further hydrogen can be removed, with the formation of acetylene. Acetylene in turn is used to produce vinyl chloride, acetaldehyde, trichloro- and tetrachloroethylene, vinyl acetate, chloroprene, acrylonitrile, etc. Hydrogen removal can also give rise to aromatics, of which benzene is particularly valuable. Benzene is converted to styrene, phenol, cyclohexane, aniline and maleic anhydride.

Oxidation

Oxidation processes can be carried out without altering the basic carbon structure of hydrocarbons. Such processes give rise to alcohols from linear chain alkanes. Alternatively, oxidation can be carried out by processes which rupture the carbon chains of hydrocarbons to give oxidative fission products. Methane can be converted under appropriate conditions to formic acid, formaldehyde or methanol. Ethane and ethylene can be converted to acetic acid or acetaldehyde, while hydrocarbons containing three carbon atoms can be converted to acetone, propionic acid, propylene oxide or n-propanol.

$$CH_4 \xrightarrow{[O]} CH_3OH \xrightarrow{[O]} HCHO$$

methane   methanol   formaldehyde

$$\downarrow {[O]}$$

$$HCOOH$$

formic acid

$$C_2H_4 \xrightarrow{[O]} CH_3CHO \xrightarrow{[O]} CH_3COOH$$

ethylene   acetaldehyde   acetic acid

$$C_3H_8 \xrightarrow{[O]} C_3H_7OH \xrightarrow{[O]} C_2H_5COOH$$

propane n-propanol   propionic acid

Chlorination

Chlorination of methane with chlorine gas at temperatures of 200°–400°C gives rise to mixtures of carbon tetrachloride, chloroform, methylene dichloride and chloromethane. Chlorination of ethane and pentane to give chloroethane and amyl chloride are very important related processes. The addition reactions of unsaturated hydrocarbons with chlorine are of key significance, as indicated above.

$$C_2H_6 + Cl_2 \longrightarrow C_2H_5Cl + HCl$$

ethane   chloroethane

$$H_2C=CH_2 + Cl_2 \longrightarrow CH_2ClCH_2Cl$$

ethylene   1,2-dichloroethane

Chlorination of high-molecular-weight hydrocarbons at about 100°C gives rise to compounds used as lube oil additives, flame retardants and plasticizers.

Other processes

The range of petrochemical processes is almost as wide as the wit of the inventive chemist and chemical engineer: examples include the manufacture of carbon black and hydrogen from high-molecular-weight hydrocarbons and the further conversion of hydrogen to ammonia.

Depending on the degree of vertical integration in a petrochemical company, the base chemicals produced on a vast scale by the processes indicated above may be sold to the chemical industry or further processed by the parent company. While speciality chemicals such as pharmaceutical chemicals are still the province of the downstream chemical industry, the oil and petrochemical companies have made major inroads into the production of bulk chemical products. An important example is the production of bulk polymers, where oil and petrochemical companies have a major stake. Although such a downstream chemical production plant may be located on the same site as the feeder oil refinery and petrochemical plant, discussion of such operations and their wastewaters is beyond the scope of this chapter.

3.1.3 Wastewaters from petrochemical works

As with oil refineries, wastewaters from petrochemical works fall into the categories of surface run-off, process-water and cooling-water. Plant design and good housekeeping to reduce water pollution from surface run-off is even more important for petrochemical plants than for oil refineries. Many petrochemical plant products, byproducts and wastes are highly toxic and polluting in water or air, and may well be highly flammable in addition. Chlorinated hydrocarbons are toxic and hazardous in terms of water pollution effects even at very low concentrations. Volatile hydrocarbons are often acutely toxic to man as well as to fish and certainly give rise to explosion and fire risks, especially in confined spaces.

Process wastewater arises from addition water used in processes as well as water produced as a byproduct in oxidation reactions. Process wastewater also arises from the washing down of process and storage tanks and pipework, as well as transport vehicle washing. Unless processes are carefully designed to minimize the generation of wastewater, a great deal of highly polluted wastewater can arise from such operations. Recent years have seen the move towards solvent cleaning of plant instead of water cleaning, with solvent redistillation and recovery to minimize loss.

Cooling-water is used very extensively in the petrochemical industry because of the high temperature reaction conditions and exothermic reactions employed. As with oil refinery processes, leaks in cooling systems give rise to significant pollution of cooling-water, which must then be treated before discharge. This applies particularly to the blowdown from recirculating cooling systems, which may be highly polluted.

Again, it is difficult to generalize about the characteristics of the wastewaters from petrochemical works because of the very wide range of possible processes in operation. The problems of water-immiscible oils and solvents, possibly high biological oxygen demand (BOD) and significant suspended solids levels are common to both oil refineries and petrochemical works. In addition, petrochemical wastewaters can contain organic compounds which are toxic to micro-organisms, ranging from phenolic compounds, through aldehydes to organochlorine compounds. Characterization of wastewater streams must be thorough to give any chance of designing an effective wastewater treatment plant. The human health and safety risks arising from wastewaters must be most carefully considered throughout.

Wastewater discharge from petrochemical works may be to sea via pipelines, into tidal reaches of estuaries, into rivers or other fresh waters or, for small-scale operations (usually), to sewers and hence to municipal sewage works. Very wide variations in final wastewater quality will be required according to the discharge route of the wastewater. Again, discharges to estuaries and coastal waters which have not been very strictly controlled in times gone past are now becoming subject to more stringent legal standards.

3.2 Environmental pollution from petrochemical works wastewaters

3.2.1 Oil pollution

Contamination of drinking-water sources
Human sensitivity to oily taints in drinking-water ranges from detection of $0.00005 \, mg \, l^{-1}$ for petrol (gasoline) containing the normal additives to $0.005 \, mg \, l^{-1}$ for diesel oil, or as much as $0.2-1.0 \, mg \, l^{-1}$ for the more inert oils. It is therefore unlikely that anyone could unwittingly drink harmful quantities of oil or petrol. On the other hand such sensitivity means that small quantities of oil can pollute very large resources of potable water. Rivers and other surface waters supplying water treatment works for potable supply are very vulnerable to contamination by oil and this occurs quite frequently in the UK, though rarely as a result of refinery operations. Oil spills onto land could eventually result in the long-term contamination of groundwater supplies. High concentrations of oil in water entering water treatment works impair filters and can cause expensive damage.

Contamination of rivers and lakes
Trace quantities of oil give rise to a visible 'rainbow' effect on the surface of water which is aesthetically very undesirable. As little as $0.15 \, \mu m$ of oil thickness on water gives colour effects. In addition to giving an unattractive appearance, a thin film of oil on water inhibits or completely stops the transfer of oxygen from the air into surface waters. Under non-turbulent conditions a few molecules thickness of oil is said to have this effect, leading to rapid oxygen depletion if the water has any significant BOD.

Not all oil floats; heavy oil may sink to the bottom of rivers and lakes, smothering benthic organisms (Nelson-Smith, 1979). Oil may coat the gills of fish and generate a physical effect of asphyxiation. Alternatively, fish may die as a result of the toxicity of the oil. Experiments indicate a great variation in toxicity from one oil to another and from one fish species to another (Klein, 1962). Any water which has been in prolonged contact with fresh crude oil or its lower fractions may have extracted significant quantities of potentially toxic hydrocarbons. Aromatic hydrocarbons, for example, are more soluble in water than is generally recognized, the solubility of benzene being about $820\,\text{mg}\,\text{l}^{-1}$ and toluene $470\,\text{mg}\,\text{l}^{-1}$ at 20°C. Benzene and toluene can be lethal to fresh-water fish in the concentration range $10-400\,\text{mg}\,\text{l}^{-1}$, according to species and the conditions of the test. Naphthenic acids are minor constituents of oil which are readily leached by water and will kill fish in the concentration range $5-120\,\text{mg}\,\text{l}^{-1}$. Furthermore, the oil may have dissolved very toxic organic compounds such as phenols or organochlorine compounds before release into the environment or subsequently.

Some minority constituents of petroleum are repulsive to fish and, while not particularly toxic, may drive fish away from their normal breeding and feeding grounds. In a salmon river, trout stream or popular coarse fishing locality this effect could bring about a serious economic or amenity loss. Oil presents a danger to swimming or predatory birds (such as swans). Aquatic mammals such as beavers, otters, musk rats and similar creatures may suffer badly.

Young fish and the smaller organisms are often more sensitive to pollutants than the adults, but are less able to escape them. Worms, snails, small crustaceans and insect larvae living on the bottom scavenge the organic wastes and detritus. If these creatures die there is no fish food and fish leave or die. Oil often damages aquatic vegetation and rotting vegetation can cause rivers to become anaerobic. Rivers may eventually become a serious fire risk due to the methane given off and floating oil and rubbish.

Oil pollution of the marine environment and coasts

Marine oil pollution arises from tankers, offshore production platforms and undersea pipelines, as well as from refinery operations and general industry. A good deal is now understood about the fate of crude oils in the marine environment as a result of the sequence of tanker disasters and offshore oil platform accidents of the 1960s and 1970s (Nounon, 1980). Although crude oils vary so much in composition, the fate of oils can be broadly classified into six stages:

- Physical processes of spreading, evaporation and dispersion
- Emulsification
- Formation of tar lumps (which are now widespread in the ocean)
- Chemical degradation
- Biodegradation
- Sedimentation

Physical processes obviously depend largely on the physical properties of the oil (density, viscosity, surface tension) and the amount of energy available in the environment (wind, waves, currents, water and air temperatures). Spreading and dispersion separate the oil into layers. The surface layer contains the lightest components from which volatile components may evaporate, while the mass of oil is dispersed vertically in the water column. A number of chemical pollutants such as detergents and organochlorines dissolve easily in the surface film and are retained and concentrated.

Emulsification gives rise to an oil-in-water emulsion on the surface and a water-in-oil 'mousse' of high-molecular-weight compounds.

Hydrocarbons not lost by physical, chemical or biological processes form virtually stable agglomerates known as tar lumps, or accumulate in living organisms.

The major portion of the hydrocarbons remaining in water and sediments is degraded by chemical and biological processes over periods ranging from weeks to years. Chemical degradation occurs mainly by photo-oxidation, initiated by primary photolysis to free radicals which react with oxygen or recombine to give alcohols, aldehydes, ketones, polymers and so on. Bacteria develop naturally to remove the more easily degraded normal alkanes, and subsequently branched and cyclic alkanes and aromatic components, under aerobic conditions.

Sedimentation is of considerable importance. The arrival of the hydrocarbon on the sea-bed is harmful to life at the bottom of the ocean. Biodegradation is very much slowed down, as is photo-oxidation. Any toxic action of the oil and its constituents is therefore prolonged. Heavy crude oils exert long-term polluting effects and are very difficult to locate or treat.

The toxicity of crude oils to marine creatures has been a subject of considerable confusion. It is now clear that many of the most toxic constituents are short-lived and, while they may be very harmful while present, they do not exert long-term effects. In general, crustaceans and certain benthic organisms, especially burrowers, are most sensitive ($1-10\,\text{mg}\,l^{-1}$), while fish and bivalves are moderately sensitive ($10-100\,\text{mg}\,l^{-1}$). Lethal concentrations may be much lower for the most sensitive larval and juvenile forms ($0.1-1\,\text{mg}\,l^{-1}$).

The sub-lethal effects of oil pollution may in fact be of greater environmental significance. Many organisms show sub-lethal effects of aromatic hydrocarbons in the parts per billion range. The effects on animals and plants include delayed cell division in phytoplankton, abnormal fish spawn, inhibited mating responses, decreased filter feeding activity of shell fish and so on.

As with fresh-water fish, mortality can result from coating of gills or digestive tracts of sea creatures. These effects are most significant along exposed shorelines populated by attached or relatively immobile species such as barnacles, mussels, limpets and snails. The effects of oil on sea birds hardly need description. In temperate and arctic climates several hundred thousand oil-coated sea birds freeze to death every year.

The tainting of edible fish and shell-fish is of considerable economic importance. This can result from exposure of these creatures to as little as one part per billion of oil, owing to the high lipid solubility of oil constituents. This effect can lead to accumulation of potentially carcinogenic polycyclic aromatic compounds in marine food chains. Large numbers of tumours have been reported in sole taken from San Francisco Bay in the vicinity of petrochemical waste disposal sites, while shell-fish contaminated by oil off the coast of Texas and Louisiana were frequently found to have neoplasms (Nounon, 1980).

Coastal oil pollution can damage shoreline vegetation and of course is very destructive of the amenity value of coasts and beaches.

Oil pollution of estuaries

Many oil terminals, refineries and petrochemical works are situated on estuaries where water movement is complex. The tides give rise to a piston-like movement of sea water into and out of the lower reaches. Plants and animals on their shores are thus subjected to strong currents in two directions, bringing marked changes in salinity as well as the normal problems of tidal immersion. A given 'block' of water may not escape to the open sea until it has oscillated many times, either accumulating or distributing widely its load of pollutants. As fresh water floats on salt, a net circulation is set up in which river water passing out to sea on the surface is balanced by a wedge of sea water moving upstream along the bottom. Estuaries are rich centres of ecological development. They are valuable as nurseries for fish taken commercially further offshore, as well as for their own potentially rich fisheries and shell-fisheries. They provide nutrients for adjoining coastal waters and feeding-grounds for many species of bird. An ecosystem of this sort is fragile, easily disturbed and recovers only with difficulty. Oil or chemical pollution can have a devastating impact on the ecology of an estuary.

The Tees estuary in the UK provides a good example of the problems of balancing the needs of industry and the desire of the local community for employment and an attractive environment. Concern at gross pollution of the 25 miles stretch of the lower Tees river and estuary began in the 1920s shortly after the development of a major chemical works on the river at Billingham. By 1937 the continued discharge of untreated industrial wastewaters had made the river impassable to migratory fish. Since 1945 four major chemical companies, an oil refinery and petrochemical works, and iron and steel works have added to the discharges into the river.

Major efforts have been made in the past decade to reduce industrial discharges. Imperial Chemical Industries (ICI), for example, has cut its daily BOD load from 480 tonnes in 1970 to 130 tonnes in 1980. The overall BOD load to the River Tees has been reduced by 70% during this period. However, the middle and lower reaches of the Tees are still classified as of 'doubtful' or 'grossly polluted' quality, and wastewater discharges into the river continue to prohibit the passage of migratory fish.

3.2.2 Chemical pollution

Petrochemical works give rise to wastewaters containing a very wide variety of organic chemicals. Some of these are largely water-immiscible and may indeed be volatile and highly flammable. Other organic chemicals may be extremely water-soluble; many are toxic towards fish or man.

Chemical pollution of drinking-water sources

In many rivers the proportion of domestic and industrial wastewater is increasing, while at the same time use is being made of lowland river waters as a source of public supplies. Inevitably the use of contaminated waters as a source of municipal supplies significantly increases the likelihood of ingestion of these chemicals, particularly as conventional water treatment is only partially effective in their removal.

The chief concern centres on the fact that some of these organic contaminants are known or suspected carcinogens. Whether their presence in minute amounts in drinking water poses any threat to human health is the subject of much current debate, particularly in the USA. In 1974 the Environmental Protection Agency (EPA) discovered 36 organic chemicals, in parts per million or parts per billion concentration, in the New Orleans water supply. The water, taken from the Mississippi River, contained several known or suspected carcinogens including benzene, carbon tetrachloride, chloroform, dichloroethyl ether and dimethylsulphoxide (McCaul and Crossland, 1974, p. 151). At the same time the Environmental Defense Fund issued a report on an epidemiological study which suggested a link between cancer deaths and drinking water obtained from the Mississippi. The large petrochemical works along the banks of the Mississippi are heavily implicated in these pollution problems. In view of the special concern about halocarbons, an analysis of drinking water from 80 separate sources was carried out looking for six low-molecular-weight halogenated hydrocarbons. Trace amounts of some of these compounds were found in all the samples, and chloroform (a carcinogen in some rodents) was found in them all, in concentrations ranging from 1 to 311 parts per billion. Some of the water samples were examined in greater detail, and 85 organic compounds were identified as being widespread contaminants. A recent EPA report to Congress listed a total of 253 organic chemicals that have been identified in drinking water.

In the UK similar concern has been expressed at the increasing number of organic chemicals that are being detected in river waters and public water supplies (Bradfield and Rees, 1978). Major studies into the health significance of these chemicals have been initiated, particularly at the laboratories of the Water Research Centre.

Apart from health hazards, petrochemical works wastewaters may give rise to problems of odour and taste in drinking water supplies. Thus the odour threshold for hydrogen sulphide is 0.0011 mg l^{-1} while that for ammonia is 0.37 mg l^{-1}. Phenolic compounds are particularly undesirable from this point of view, since water chlorination for disinfection can give rise to chlorophenols with extremely low thresholds of olfactory detection in drinking water (0.000 18 mg l^{-1} for chlorophenol).

Chemical pollution of rivers and lakes

Many of the organic chemicals present in petrochemical works wastewaters are biodegradable. Without treatment, such wastewaters and chemicals discharged to a river or lake give rise to high BOD. The consequent oxygen depletion causes fish death and ultimately anaerobic conditions.

On the other hand, some organic chemicals present are biocides, including phenols, aldehydes and chlorinated hydrocarbons. Many such compounds are very toxic to fish (Table 3.1) and may break down only slowly in surface waters,

Table 3.1 Toxicity of some organic chemicals to fish (Klein, 1962)

Chemical tested	Fish species	Lethal concentration LC_{50} (mg l^{-1})	Exposure time (h)
Amyl alcohol	Goldfish	1	161
Aniline	Brown trout	100	48
DDT	Rainbow trout	0.5	24
Dieldrin	Rainbow trout	0.05	24
Naphthalene	Perch	Chemical	1
Para-cresol	Rainbow trout	6	2
Phenol	Rainbow trout	6	1
Quinoline	Perch	30	1

particularly if they are water-immiscible and sink to the bottom of rivers and lakes.

An important danger is the possibility that chemicals of relatively low toxicity to fish, but greater persistence in the environment, will bioaccumulate through aquatic food chains. Among the organic chemicals certain chlorinated hydrocarbons have proved a particular problem in this respect. Fish-eating birds of prey have been severely damaged by ingestion of fish contaminated with a variety of persistent organochlorine compounds. Man has not been immune from such effects; polychlorinated biphenyl levels in fish caught off the coast of Japan have given rise to particular concern. As with oil pollution, more recent research on the effects of persistent toxic organic chemicals on ecosystems has concentrated on sub-lethal effects, including reproduction difficulties of birds and fish and behavioural effects.

While the most obvious problems of oil refinery and petrochemical works wastewaters in surface waters stem from concern about organic chemicals present, the effects of inorganic chemicals must not be overlooked. Clearly, large fluctuations in pH of effluents discharged to rivers and lakes are highly undesirable, while certain inorganic chemicals such as ammonia are very toxic to fish. Oil refineries generate very large volumes of salty water which can well turn fresh-water rivers and lakes brackish if discharged untreated. In most countries rivers and lakes are utilized for many purposes other than provision of drinking-water for people. These include water for agricultural use (such as drinking-water for animals or irrigation of crops), industrial water needs, navigation and recreation. Pollution of rivers and lakes by organic or inorganic chemicals from oil refineries and petrochemical works inhibits or prohibits their use for some or all of these purposes.

Chemical hazards to sewers and sewage works

Oil refinery and petrochemical works wastewaters are not commonly discharged to sewers running to municipal sewage works, because of the very large volumes involved. However, if a chemical works dealing with smaller quantities of petrochemicals is located conveniently to a sewage works this may be an attractive option.

As with all discharges to sewer, it is important to protect sewer workers, the fabric and operation of the sewer and the sewage works, the workers at the sewage works, and also the river to which the sewage works discharges its treated wastewater. The main potential hazard to sewer workers and sewage works personnel arises from the dissolved gases and volatile solvents which may be present in petrochemical works wastewaters. These can give rise to flammability and even explosion risks, and to both acute and long-term toxicity problems. Some solvents, while not necessarily very toxic, may exert an anaesthetic effect and cause sewer workers to slump unconscious and drown in sewage.

Numerous classes of organic compounds have been found to interfere with the processes of biological treatment in sewage works. Organochlorine compounds, in particular, frequently give rise to problems (Table 3.2). Fortunately, some of the most common polluting chemicals from this viewpoint are volatile compounds which are lost by evaporation at an early stage in treatment.

A more subtle problem arising from discharge of organic chemicals in

Table 3.2 Concentrations at which some organic chemicals will inhibit anaerobic digestion of sewage sludge (Bailey, 1976)

Chemical	Toxic concentration in sewage ($mg\,l^{-1}$)	Toxic concentration in sludge ($mg\,l^{-1}$)
Benzene		50–200
Carbon tetrachloride		10
Chloroform		0–1
Methylene chloride	1	
Pentachlorophenol	0.4	
Tetrachloroethylene		20
Toluene		430–860
1,1,1-Trichloroethane		1
Trichloromethane	0.7	

industrial effluents to sewage works is the possibility that some will pass through the works without significant separation or biodegradation, going on to pollute surface waters. A survey and desk study of industrial wastewater discharges to sewage works along the River Lee, near London, has shown that 11 organic compounds were of concern. These were deemed to be non-biodegradable and could enter the river and subsequently be withdrawn into water intended for treatment for drinking-water supplies. These compounds included o-phenyl phenol, $2,2'$-dihydroxy-$5,5'$-dichlorophenylmethane and pentachlorophenol. River water samples have been analyzed and fairly sound agreement has been established between the levels deduced and those found in practice (Wood and Richardson, 1978). Expert medical opinion is being sought on the effects of ingestion of these compounds. A similar exercise is being undertaken for chemicals being discharged to the non-tidal Thames and its tributaries. Such studies indicate that for certain organic chemical pollutants in wastewaters conventional sewage works treatment is not sufficient to ensure complete removal.

Sea discharge of chemical wastewaters

Since many oil refineries and petrochemical works are located on estuaries and coasts, disposal of wastewaters by discharge to sea is an attractive option on economic grounds. While discharge of wastewaters via short pipelines gives rise to pollution of the shoreline and bathing waters, discharge via long pipelines under appropriate tidal conditions may be effective in sending wastes far out to sea. In addition to the great dilution of wastewaters in sea water, which limits their polluting effects, the sea has a great capacity for treating certain types of wastes. Thus, the buffer capacity of the sea copes with large volumes of extremely acid or alkaline wastewaters. The wave action of the sea introduces oxygen readily into the surface water and makes oxygen depletion unlikely. Biodegradable organic compounds break down readily in the open sea provided that they do not sink into bottom sediments.

Clearly, the sea cannot be regarded as a sink for all wastes. As discussed above, oil pollution of the sea is a serious problem. Similarly, pollution of coastal waters by persistent organic compounds such as organochlorine compounds, and toxic inorganic metals such as mercury and lead, is a growing threat. The danger is particularly acute where large volumes of polluting wastewaters are discharged into tideless seas or into coastal waters where the prevailing tides do not allow effective dispersion of the wastes.

Some of the hazards of discharging wastewaters from petrochemical works into estuaries have been discussed above. In addition to their damage to the ecosystem of the estuary, the pollutants which find their way to the estuary mouth may be deposited there and give rise to further damage. The mouth of an estuary is a particularly important ecological environment. Fish will thrive on the food sources provided by estuarine waters, the bottom of the estuary mouth being particularly fertile. If persistent toxic chemicals are deposited here they may well be taken up by bioaccumulation and cause long-term damage to important fisheries.

3.2.3 Physical forms of pollution

Problems of physical pollution by floating, immiscible liquids from petrochemical works are similar to those discussed under oil pollution. Coloured effluents add an extra dimension to the objectionable aesthetic effects of industrial water pollution. However, the simple chemicals produced by petrochemical works are normally colourless.

Depending on the nature of the processes involved, petrochemical works give rise to significant levels of suspended solids in wastewaters. These are likely to be contaminated with oil and, if discharged to surface waters, give rise to oil pollution in addition to the more expected problems of suspended solids in surface waters. Suspended solids with oil droplets attached may neither sink nor float in effluent streams and are a particular difficulty in treating oil refinery effluents. Petrochemical works alone are less likely to give rise to high levels of suspended solids in wastewaters than are oil refineries, since both the raw materials and the products are usually volatile liquids and gases.

Certain processes give rise to still residues of tars and polymers which are highly polluting. These residues should be so processed that they do not find their way into the main plant wastewater streams but are separated as special wastes.

For discharges of wastewaters to surface waters or sewers, thermal pollution is perhaps the most potentially troublesome of the physical forms of pollution. As described above, petrochemical works use vast amounts of cooling-water. If heated water from cooling plants is discharged into watercourses, particularly rivers and lakes, a number of undesirable consequences may ensue. The water temperature rise itself may lead to changes in the ecology of rivers and lakes. Wide fluctuations in temperature are usually harmful. More harmful in already polluted rivers is the release of oxygen from the water as temperatures rise. If the water is polluted and the oxygen level already low, a further reduction due to thermal effects may lead to fish deaths. In addition, normal processes of biodegradation of pollutants are speeded up at higher temperatures and this in turn leads to more rapid oxygen depletion and the onset of anaerobic conditions.

Undoubtedly, one of the advantages of discharging hot wastewaters into the sea is the vast cooling effect of sea water. Discharge of hot wastewaters into estuaries, on the other hand, may not have the expected results because of the complexity of the environment, and may do further harm to the delicate balance of nature.

Discharge of hot wastewaters into sewers is undesirable in view of the hazards to sewer workers (particularly, the volatilization of dissolved toxic gases in sewage) and the acceleration of destructive chemical and biological processes which damage the sewer.

3.3 Legal constraints on wastewater discharges

Every country has its own, often very particular, network of laws, regulations and institutions designed to control discharges of wastewater from petrochemical works. No attempt can therefore be made to give a full account of legislation on a world-wide basis. Reference is made to attempts at legislative controls on a European regional basis and, by way of example, some detailed discussion of UK national legislation places subsequent discussion of standards in context.

3.3.1 European regional legislation

The Paris Convention for the Prevention of Marine Pollution from Land-based Sources, 1974, has been adopted by ten Western European states. The parties to the convention have agreed to combat marine pollution from land-based sources. This includes pollution via watercourses, and from the coasts, including discharges from pipelines. The parties have agreed to eliminate pollution by so-called 'black list' substances, and strictly to limit pollution by 'grey list' substances. The 'black list' includes 'persistent oils and hydrocarbons of petroleum origin', and the 'grey list' includes 'non-persistent oils and hydrocarbons of petroleum origin'. Persistent organochlorine compounds which may occur in some petrochemical works wastewaters are also blacklisted. Each party has agreed to take any necessary steps in its territory to prevent and punish conduct which contravenes the convention.

The United Nations Environment Programme aims to have regional Action Plans adopted for seven seas around the world. It is likely that the Action Plans will include conventions, similar to the Paris convention, dealing with land-based sources of pollution. The European Economic Community (EEC) is also very active in environmental affairs and has already issued a number of environmental directives relating to industrial effluent discharges to watercourses. Of particular relevance to oil refineries and petrochemical works are the Directive on the Disposal of Waste Oils of 1975, and the Directive on the Protection of Groundwater against Pollution (forthcoming). The Directive on the Quality of Surface Water Intended for the Abstraction of Drinking Water and the Directive on Drinking Water Quality may indirectly affect petrochemical works discharging wastewaters into sources of supply.

With all these international regulations, however, there is doubt about the effectiveness of enforcement in the absence of strong political pressure and an international inspection and enforcement agency. Thus, Italy has recently been brought before the European Courts of Justice for failing to adopt five environmental directives within the prescribed time limits (including several of those listed above). As far as oil pollution is concerned, many scientists and industrialists are sceptical about the effectiveness of stricter controls on land-based sources of pollution while controls on oil pollution from shipping remain weak and ineffectual.

3.3.2 United Kingdom legislation

From the UK experience of water pollution control over many years, perhaps the most effective means of preventing pollution by particular noxious materials is to make their unlicensed discharge an offence, but to provide for discharges with the consent of a licensing authority. These discharges are then subject to the conditions imposed by the authority, and to regular inspection. This system is essentially the one operated by the water authorities in England and Wales, and by the river boards in Scotland. Discharge to sewers and sewage works is controlled under the Public Health Acts of 1936 and 1937. Discharge to rivers is controlled under the Salmon and Freshwater Fisheries Act of 1923 and the Rivers (Prevention of Pollution) Acts of 1951 and 1961. Discharge to underground water is controlled under the Water Resources Act 1963, and to estuaries under the Clean Rivers (Estuaries and Tidal Waters) Act 1960. The consent conditions for discharge of polluting materials in wastewaters have been based on full consideration of the nature of the receiving environment and its ability to cope with the pollutants without significant deterioration. This approach is maintained in Part II of the Control of Pollution Act 1974, which has drawn together the main features of water pollution control legislation in the UK.

In practice, standards have varied widely from one water authority to another, even for discharges of apparently similar wastewaters to similar watercourses. Some discharge consent limits seem to have been set on an arbitrary, or rule-of-thumb, basis, rather than on any scientific criterion. Discharges to estuaries and tidal waters have often been dealt with on a particularly arbitrary basis, strict limits being set in some instances and virtually no limits in others. In an attempt to achieve scientifically based standards, the water authorities have been going through an exercise of designating target water quality standards for inland watercourses and relating industrial wastewater discharge consent conditions to these targets (National Water Council, 1978). In view of the increasing international concern about discharges to the marine environment, there will be increasing efforts to designate more detailed water quality standards for estuaries and coastal waters. Again, industrial wastewater discharge consent conditions will inevitably be reviewed in the light of these standards. A tightening of standards would have a very considerable impact on UK petrochemical works discharges.

Unfortunately, the UK and the other EEC countries are in disagreement over the philosophy of controlling wastewater discharges. The EEC Commission and the other Member States desire a system with uniform discharge standards for particular types and sizes of industrial works. These standards would take no account of the nature of the receiving watercourse. While this approach has merit with reference to the discharge of industrial wastewaters to long inland rivers such as the Rhine, or to landlocked seas such as the Mediterranean or the Baltic, it is not sensibly applicable to Britain. Already there have been protracted negotiations over discharges from the paper and pulp industry and the titanium dioxide industry, with the UK arguing that wastewater discharge limits for discharges to the North Sea or the Atlantic need not be so strict as for discharges to the Rhine or to the Mediterranean.

3.4 Standards for discharge

3.4.1 General principles

From the above discussion of legislation there are clearly at least two quite separate approaches currently adopted in Europe alone with respect to standards of wastewater discharges. With the flexible UK approach, based on water quality standards, the consent conditions for wastewater discharges will vary greatly according to the nature of the receiving water environment. Standards will become tighter in the approximate order of discharge to the sea, to estuaries, to municipal sewers, to polluted lowland rivers, and to clean upland rivers (or other clean surface waters).

According to what the works is producing, the nature of the wastewater or wastewaters is very variable, and it is difficult to envisage how the EEC Commission approach of rigid emission standards could be applied, particularly to petrochemical works.

Normally, untreated wastewaters might be expected to contain free floating oil, possibly emulsified oil, suspended solids, biodegradable organic materials, toxic and non-biodegradable organic compounds, ammonia and inorganic compounds including traces of heavy metal catalysts and salinity, with variation of pH and temperature.

The discharge consent conditions set by the water authority will depend on a number of factors including some that are unquantifiable, such as the past history of the region, including incidents of pollution, the length of time oil refineries or petrochemical works have been located in the region, and the degree of concern about environmental matters in the region generally, and particularly near the actual location of the works. Clearly, the nature of the receiving water and the uses to which it is put will be of particular significance. Considerations with regard to quality of life have led to a general raising of standards, as have increasing knowledge and concern about the effects of water pollution on human health and safety.

Recent years have seen an increasing emphasis on controls to limit the polluting effects of accidental spills and the careless disposal of wastes from works, including wastes from wastewater treatment plants themselves.

3.4.2 Wastewater discharge standards for oil

For oil refinery and petrochemical works, one of the most difficult aspects of trying to set standards for effluent quality concerns the definition of 'oil'. Since the material is so complex, any definition can only be a compromise. Thus, should hydrocarbon gases and volatile components that weather rapidly be included in the analysis and therefore the definition of oil-in-water discharges? The *Shorter Oxford English Dictionary* defines oil as '... a liquid at ordinary temperatures, of a viscid consistency and characteristic unctuous feel, lighter than water and insoluble in it'. For discharges to rivers and sewers, however, the water authorities may specify that the standard applies not just to immiscible oil, but also to oil in emulsified form, or even to all oil and related substances which can be extracted from the aqueous phase of the wastewater sample with the aid of a specified water-immiscible solvent. For example the standard may refer to 'non-volatile ether-extractable matter'.

Sampling of oil in wastewaters presents difficulties. Oil in the wastewater can exist as a separate heterogeneous phase, or as a dissolved or emulsified phase in the body of the water. Both phases can occur together, but seldom is the insoluble oil uniformly distributed throughout the total wastewaters. Representative sampling of oily wastewaters can be extremely difficult. In view of the degree of pollution possible from small quantities of oil, and the difficulties of definition, sampling and analysis, standards for river discharge may be extremely strict. Limits such as 'no visible trace of oil on the receiving water', or 'no physically separate oil' or simply 'no oil' are typically set by water authorities in Britain (Fielding, 1976).

Small quantities of free or emulsified oils may be allowed into sewers feeding sewage works. This will depend on the nature of the oil, other constituents of the wastewater, and the capacity of the sewage works to cope with small quantities of oil. In addition to concern about the effects of oil on the functioning of the sewage works, the relevant authority will be extremely concerned to ensure that no flammability or explosion hazard exists. Petroleum or mixtures containing petroleum or other volatile flammable liquids must be prevented from entering sewers.

Standards for discharge to estuaries or coastal wastes from pipelines have been generally less severe, with concentrations of between 5 and 25 mg l^{-1} total oil being quoted according to local circumstances. Often low limits for industrial wastewater discharges are negated in terms of environmental improvement by the much greater oil discharges from shipping, both at sea and in harbours. Legislation on oil in the marine environment in the past has been directed at the avoidance of fouling the surface of the sea: it has been largely concerned with immiscible, non-volatile oil. Thus, analytical methods specified for oil-in-water standards employ physical separation methods which do not measure the soluble components of oil, such as light aromatics, naphthenic acids, various phenols and so on. This concept has been justified in terms of the ready destruction of these components by the bacteria present in the sea, while the heavier aromatics remain in the objectionable immiscible oil phase (Oil Industry Forum, 1979). However, the toxic effects of the water-miscible oil fractions described cannot be neglected, particularly in the context of estuarine and coastal pollution.

3.4.3 Standards for other constituents

Discharge to non-tidal rivers and lakes

Historically in Britain, industrial wastewater discharges to rivers have been controlled by standards relating to those applied to sewage works discharges. Typically a BOD_5 limit of $20\,mg\,l^{-1}$ and a suspended solids limit of $30\,mg\,l^{-1}$ have been applied to sewage works wastewaters discharging into a river giving at least 8:1 dilution of the wastewater. Higher standards are often required if dilution is less, or particularly if the river water is subsequently abstracted for potable supply. In more recent years the BOD_5 standard has been largely replaced by a chemical oxygen demand (COD) standard. This has had an impact on petrochemical works wastewaters, where a significant proportion of the organic compounds present are only slowly biodegradable and were therefore not measured on the five-day BOD test.

A wide range of other consent limits are placed on industrial wastewaters discharged to rivers; typical values are given in Table 3.3. These consent limits

Table 3.3 Typical consent conditions for discharge to river of certain pollutants in industrial effluents (concentrations in $mg\,l^{-1}$)

Parameter	Discharge limit
Temperature	30°C
pH	pH 5–9
Suspended solids	30
BOD_5	20
Ammonia (as N)	10
Sulphide (as S)	0.5
Cyanide (as CN^-)	0.1
Chlorine	0.5
Toxic heavy metals (total)	1.0
Mercury	0.01
Cadmium	0.05
Lead	0.1
Phenols	0.5
Oil and grease	None visible
Individual toxic organics as specified	Very low values or none

reflect common river pollution problems with industrial wastewaters. Outside a small group of known harmful classes of chemicals, such as oils, detergents, phenols and organochlorine compounds, insufficient consideration has been given to identifying and limiting specific organic pollutants unless they were known to be present and toxic (for example, pesticide residues). In view of concern about traces of persistent organic substances in surface waters for potable supply, and in fish caught for human consumption, it is to be expected that much more attention will be paid to organic compounds in industrial wastewaters in the future. Compounds which persist in the environment (or which break down to give other compounds which persist), which may be harmful in the long term, and particularly those which build up through food chains, will have very strict discharge consent limits applied. The insidious effects of such compounds, which have been highlighted by environmental and human disasters in Japan particularly, are now recognized world-wide. Such increasing concern about harmful effects of organic or organo-metallic compounds will be reflected in rising standards for petrochemical works wastewaters in many countries.

These developments show that the treatment of petrochemical wastewaters for discharge to rivers and lakes is likely to become even more complex and expensive in the future.

Discharge to coastal waters

Standards for discharge of wastewaters from petrochemical works into estuaries and coastal waters are particularly important since so much wastewater is disposed of by this route. As discussed earlier, the sea has a large capacity for absorbing and dealing with certain pollutants without undergoing any deterioration. This applies particularly around the coast of Britain, where strong tides and the vigorous scouring action of the sea disperse and destroy wastes. Thus an attractive option for discharge of petrochemical works wastewaters is by long pipeline into the sea. With correct siting and operation, very rapid dilution of wastewaters can occur and natural treatment may be rapid.

Volume and temperature levels are unlikely to be significant except in the case of very large discharges of hot fresh water. This is unlikely to arise in wastewaters from petrochemical works today, if only because of energy conservation considerations. The sea has an almost infinite buffer capacity for acidity or alkalinity in a wastewater. Provided that an immediate dilution of at least 100-fold is achieved, wastewaters of pH as low as 1 or as high as 13 can be safely discharged without any toxic effect. Similarly, dissolved salts are most unlikely to affect the sea unless any constituent is toxic (as discussed below). For example, an immediate dilution of 10 times renders a 35% saturated brine solution completely compatible with sea water at 3–5% solids.

While some non-toxic suspended solids may be discharged without significant effect, dense suspended solids can precipitate in the sea and accumulate at the outfall, blanketing the sea bed. Consent limits therefore clearly reflect the existing environmental situation and the effects of tides.

There is a distinction between a muddy estuary with high natural silt and a clear-water rocky bottom. Conventionally, maximum levels in the wastewater are set at 150–300 mg l^{-1}, but higher levels could produce no ill effect in some circumstances. Of course, solids that are contaminated with oil will give rise to oil pollution and must be strictly controlled at source.

As discussed earlier, oxygen demand (whether BOD or COD) is another property for which tidal waters can provide effective treatment by a combination of biological and physicochemical reaction. The principle is to avoid reduction of the dissolved oxygen that could affect marine life. It is practically impossible to deoxygenate the open sea around the coast of Britain because of the high oxygen replenishment due to tides and wave action. Even with moderate dispersion, BOD loads of several hundred tonnes per day have no effect upon the open sea. This argument has much less validity for essentially landlocked or tideless marine environments where wave action is low.

Non-persistent toxic materials which can be rapidly diluted to concentrations below the 96 hours LC_{50} (concentration lethal to 50% of the test creatures) for critical species will not give rise to serious environmental problems. Compounds which are extremely toxic even at very low concentrations should never be discharged in wastewaters whether to sea or elsewhere; persistent toxic materials require careful consideration. The criteria are the rapid dilution and dispersal of wastewater below the toxic concentration to avoid any sub-acute toxicity effects in the area, and the elimination of any concentration by bio-accumulation through food chains. Bio-accumulation can be studied directly on critical species such as shell-fish, and appropriate levels set for any relevant materials in the wastewater. Unfortunately, sub-lethal effects are more difficult to allow for because little is known about them. Effective control can only be achieved by regular ecological monitoring of the area.

Clearly, discharge of oil refinery and petrochemical wastewaters to sea offers the possibility of effective pollution control without such heavy investment in treatment plant as is required for river discharge. However, it is by no means necessarily a cheap or easy option. Thus, ICI has recently invested in a scheme to collect wastewaters from its chemical manufacturing site at Stevenson, Ayrshire, Scotland, a number of which currently discharge to the River Garnock, and transfer them to a pumping station near the shore. From there the wastewater will be transferred to deep sea water via a submarine pipeline more than 1.6 km long. Extensive tests by ICI Marine Biology Laboratories at Brixham have been carried out with the Clyde River Purification Board to ensure that the scheme will present no risk to the submarine environment. The total cost of the scheme is in the region of £10 million.

In all such schemes designed to take advantage of the treatment and purification capacity of the natural environment, close environmental monitoring is essential to ensure that pollution control does not become semi-controlled pollution.

Discharge to estuaries
Since estuaries are tidal and contain saline water it is very tempting to apply the same arguments for discharge of only partially treated wastewaters as can be applied to sea discharge. However, as described above, estuaries are hydrologically and ecologically complex. Many estuaries have suffered considerable abuse due to lack of understanding of these complexities. Great care is necessary in setting consent conditions for wastewater discharge. In many cases

the discharge consent standards need to be much closer to the standards for river discharge than to the standards for discharge into the sea. Thus, while it is practically impossible to deoxygenate the open sea by discharge of high BOD wastewater, it is perfectly possible to deoxygenate an estuary. A number of smaller estuaries and tidal waterways around the coast of Britain have suffered for many years from severe oxygen depletion. Similarly, great care is needed when considering discharge of toxic organic substances into estuaries, since mixing rates may be slow, and estuaries are important ecologically.

Again, a thorough programme of ecological monitoring is essential where new petrochemical works are to be sited on estuaries. For existing plants it is likely that the near future will see an increase in standards for wastewater discharges into estuaries, both in Europe and in other parts of the world.

Discharge to sewer

Sewage works are designed to deal with suspended solids and biodegradable organic materials. Thus, where the sewers and sewage works have the hydraulic and treatment capacity, some refinery and petrochemical works wastewaters could in principle be discharged to sewer, provided that oil and flammable compounds were absent and that the substances in the wastewater did not interfere with biological treatment processes at the sewage works. Typical consent conditions for discharge of industrial wastewaters to sewer are given in Table 3.4.

Table 3.4 Typical consent conditions for discharge to sewer of certain pollutants in industrial effluents (concentrations in $mg\,l^{-1}$)

Parameter	Discharge limit
Temperature	< 40°C
pH	pH 5–10
Suspended solids	400
HCN (or compounds producing HCN on acidification)	5
Sulphur compounds which on acidification liberate H_2S	10
Free chlorine	100
Toxic heavy metals (lower values for mercury, cadium and lead)	5
Synthetic detergents	30
Free or emulsified grease or oil (including all hydrocarbon fractions boiling at 150°–300°C)	Variable, ≤ 100
Volatile petroleum products and flammable solvents	None
Phenols	Variable
Tar or tar oils	Variable, low
Organosulphur compounds	Variable, low
Organohalogen compounds	Variable, low
Organophosphorus compounds	Variable, low

In practice, petrochemical works give rise to very large volumes of wastewaters, which only the largest municipal sewage works could accommodate. In addition, some of the constituents of the wastewater (such as phenols, aldehydes and mercaptans) are likely to interfere with biological treatment. There is also the continuing possibility of spills or other accidents releasing quantities of oil and flammable and toxic compounds into the sewer. In addition, the charges made by water authorities for discharge of high BOD/COD wastewater to sewage works is very high.

On the other hand, this option may be more attractive than direct river discharge for an inland works. The sewer system and sewage works does provide some buffer capacity against accidents, compared with direct discharge to river. Some industrial effluents can also be better treated in a mixture with domestic sewage. Some works may not be conveniently sited for the inclusion of a full wastewater treatment plant on the site.

A compromise worth considering is the use of a partial treatment plant on the works site which removes harmful constituents and carries out partial biological treatment, the partially treated wastewater being discharged to sewer for final treatment at the sewage works. This may prove most cost-effective, and also ensures that any accidents will affect the works wastewater treatment plant before damaging the municipal sewage works.

3.5 The treatment of petrochemical works wastewaters

3.5.1 Treatment principles

The increasing costs of oil and chemicals, together with problems of availability and costs of water of various qualities, have led to greater concern about materials management and waste reduction in the petrochemical industry. In addition, the costs of wastewater treatment (and cost of disposal of wastes arising from treatment) have risen rapidly as environmental standards have become more exacting.

Considerable success has been achieved over recent years in reducing the polluting load of petrochemical plant wastewaters and hence the cost of treatment. As stressed above, a major change has been the movement away from the use of chemicals in processing. Water is best conserved by recycling and reusing it on the basis of a hierarchy of quality, with demineralized water and condensate at the top of the hierarchy, town's water and raw water next and cooling-water at the bottom. Used water is recycled within the original process if possible, or passed to the next lowest level in the hierarchy which is able to accept it. Only as a last resort should used water be discharged to drain as a wastewater.

Overall design of plants to achieve good housekeeping also has an important role to play. Thus, the roofing of surface areas where oil spills are likely or inevitable will reduce the amount of oil getting into surface run-off from rain. Appropriately placed bund walls, sumps and oil-traps can collect and contain leaks and spills of oil and water-immiscible chemicals, and avoid heavy slugs of such substances finding their way to wastewater treatment plants.

Highly concentrated wastewaters, especially those containing materials which can be recovered and re-used, or those containing toxic constituents, are often best dealt with as discrete wastes, rather than being discharged to the effluent treatment plant. In this way substantial reductions in costs can be achieved, and interference of immiscible or toxic substances with smooth operation of wastewater treatment can be minimized. Certainly, where there is a good possibility of economic recovery of materials from process wastewater streams, this should be closely considered at the design stage. Once wastewaters from different processes have been mixed together, the possibility of economical recovery of any constituent is very much reduced.

Concentrated wastewater that does not contain materials worth recovering may be disposed of by incineration, or by tipping under appropriate conditions. This is likely to prove a more cost-effective approach to treatment than discharge to wastewater treatment plants, especially where biological treatment is involved.

The process wastewater and final raw wastewater from a petrochemical plant are critically dependent on the nature of the processes operated. Processes may give rise to wastewater streams containing immiscible liquids and solids (including oil), very high BOD organic loadings, volatile, flammable or toxic compounds, non-biodegradable compounds, variable acidity and alkalinity and so on. The variability of wastewaters from particular processes is shown in Table 3.5.

Table 3.5 Characteristics of effluents from particular production processes of a petrochemical plant (American Petroleum Institute, 1969)

Process/product	Capacity (t year^{-1})	Flow (m^3 t^{-1})	BOD (kg t^{-1})	BOD average (mg l^{-1})
Ethylene oxide	50 000	0.476	0.88	1 800
Isopropyl alcohol	31 500	2.540	0.99	400
Butadiene	33 000	1.695	0.63	400
Phenol +	40 000	0.559	8.00	17 300
Acetone	25 000			
Acrylonitrile	38 000	4.47	38.70	86 000
Cyclohexane	0	0	0	0

Wherever possible at the design stage for new plants, or subsequently, every effort should be made to consider the recovery of worthwhile materials from petrochemical wastewaters. Many materials present are volatile and may be removed from wastewater streams by distillation, possibly under reduced pressure. While such processes are energy-intensive, the recovery of valuable materials and reduction in wastewater treatment costs will often justify the capital and operating costs of recovery plants. Less volatile solvents and chemicals may be stripped out of wastewaters by steam distillation. A wide range of purpose-designed solvent recovery units are now available and can be added to existing plants, provided that the process wastewater can be segregated for treatment.

3.5.2 Wastewater monitoring

The importance of sampling and analyzing wastewaters on a continuing basis cannot be overstressed. Purely from the point of view of design and operation of wastewater treatment plants, full data on the anticipated or actual wastewater streams are of crucial importance. In addition, wastewater monitoring gives important information about the process plant while it is operational, particularly early warning of abnormal losses. Such losses can arise from a number of insidious causes, such as poor liquid–liquid separation, entrainment from a condenser into a vacuum ejector on a still, or malfunction of equipment as a consequence of instrument error.

It is essential to consider flow measurement, which may be difficult to carry out in petrochemical works. Drains are seldom laid out to facilitate monitoring. Not only may they combine the wastewaters from functionally unrelated plants, or split the discharge of a single process, but drains may be several metres below ground level. This can hamper the construction and maintenance access of any flow measurement device that has to be located in such a situation. Flumes and weirs have often been found to be acceptable flow measurement devices for drains and conduits on petrochemical plants, although devices such as electromagnetic flow meters are becoming more widely used. Flow measurement equipment must be robust to withstand damage and clogging by a wide range of debris. Design data should be helpful in indicating the expected range of flow rates under normal operating conditions, and radioactive or chemical tracer techniques can be employed for single measurements where continual measurement is unnecessary or impracticable.

In a large works it is essential that automatic sampling is employed on a 24 hour basis. The nature of the process and the wastewater treatment plant determine where best to sample, how frequently to sample, and whether the sampling frequency should be proportional to flow rate. More frequent sampling is required where processes produce high concentrations of pollutants intermittently, while flow proportional sampling is necessary where flow rates vary widely. The extent of wastewater sampling and analysis across a petrochemical works must be judged on a cost-effectiveness basis. Full sampling at every point of wastewater production is not practicable, if only because of the maintenance effort required to keep sampling equipment operational. At appropriate points in drains detectors for flammable vapours should be installed, since these constitute a serious hazard. All monitoring equipment for use in potentially flammable environments should be designed and approved for this purpose, and systems of work for wastewater sampling and analysis must take full account of health and safety hazards.

3.6 Physicochemical treatment processes

3.6.1 Steam distillation
Steam distillation (or stripping) is most commonly used to remove water-soluble volatile components from wastewater streams. In the past the process was applied to a restricted range of simple volatile components such as

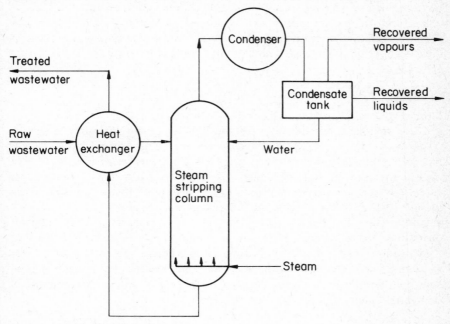

Fig. 3.1 Steam distillation plant

methanol or ammonia, but today a wide range of compounds is recovered by this method including phenol and many organochlorine compounds.

The process is usually carried out in a continuous manner employing plant similar to that used for fractional distillation (Fig. 3.1). The wastewater is preheated by a heat exchanger and is introduced into the distillation column near the top, while the steam is introduced near the bottom of the column. As the wastewater passes down the column under the action of gravity, it comes in contact with the steam and vapours rising from the bottom of the column. As the wastewater passes down the column it loses progressively more of its volatile constituents until, near the bottom of the column, it is heated by the incoming steam and is depleted as far as possible. The volatile constituents driven out of the wastewater reach the top of the column and can be taken away for processing. The hot processed wastewater from the column is taken through a heat exchanger to preheat the incoming untreated wastewater stream, thus minimizing heat consumption in the steam stripping process.

The volatile materials recovered by steam stripping are usually processed further for reuse or incinerated. If such volatiles contain sulphur or halogens the production of acid gases in the incineration process must be taken into account.

The steam stripping process is very reliable and can cope with wide fluctuations in the flow rate of the wastewater to be processed and in the concentration of volatile constituents. For chlorinated hydrocarbons such as chloroform, 1,2-dichloroethane and tetrachloroethylene well over 95% removal can be achieved from incoming wastewater streams containing in excess of 16 000 mg l^{-1}. In some cases complete removal of the volatile material can be achieved.

3.6.2 Treatment to remove less volatile constituents

For less volatile immiscible solvents and liquids, gravity flotation and similar techniques may be applied. Such techniques are discussed later in connection with oil removal from wastewaters. Tilted plate separators are now being applied with success to a wide range of solvent recovery operations. In such cases it is very important to ensure the compatibility of the plate materials (casing and accessories) with the wastewater constituents, which may have corrosive characteristics, especially when hot. For immiscible solids a range of techniques for solid–liquid separation may be appropriate, although those which can be readily applied on a continuous basis will be favoured.

Chemical constituents of value that are in solution and are not volatile may be separated and recovered by means of either solvent extraction or activated carbon treatment.

Solvent extraction

Liquid–liquid solvent extraction involves bringing the wastewater stream in contact with a water-immiscible solvent. Any organic constituent dissolved in the wastewater will partition itself between the two phases, and the solvent is chosen to have the potential to extract most of the desired material from the wastewater. The use of multiple contactors may enhance the degree of separation achieved. The process is shown in simplified form in Fig. 3.2. Depending on the ease of extraction and the percentage removal required, the plant might consist of a single-stage mixing and settling unit, several such mixers and settlers in series, or a multi-stage unit within one device operating by counter-current flows. The extracting solvent containing the extracted solute is taken off for processing. In nearly all cases such processing involves the recovery of the solvent, for example by distillation, so that it can be used again in the extraction process. Solvents are now too valuable to be used only once. Indeed, the use of solvent extraction processes can often only be justified on economic grounds where a very high percentage of the solvent used in each cycle can be recovered for reuse.

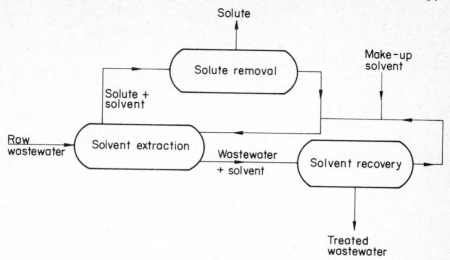

Fig. 3.2 Solvent extraction plant

In most cases the solvent used in the extraction process has itself some solubility in water. If the dissolved solvent will itself cause difficulties in further treatment (or final discharge) of the wastewater, or if significant solvent losses can occur by this route, the solvent may be removed from the wastewater in another step (for example, steam stripping).

Solvent extraction is relatively expensive and is used to treat selected concentrated wastewater streams, especially where material recovery is possible. The value of the recovered material will then offset the process costs, assuming recovery is from a segregated wastewater stream. In general where volatile solutes are present in wastewaters in moderate to low concentrations it is more economical to use steam stripping rather than solvent extraction. The principal application has been in the removal of phenol and related compounds from wastewaters. Solvent extraction can reduce phenol concentrations from levels of several per cent down to levels of a few parts per million. Removal efficiencies of 90–98% are possible in some applications and with special equipment such as centrifugal and rotating disc contactors removal efficiencies around 99% have been achieved (EPA, 1980).

The most common solvents are light oil, benzene and toluene; more selective solvents include isopropyl ether, tricresyl phosphate, methyl isobutyl ketone, methylene chloride and butyl acetate. Extraction with light oil may be followed by phenol recovery via extraction of the oil with caustic soda solution, and the phenol recovered as sodium phenolate.

Solvent extraction has been used for the removal of salicylic and other hydroxy–aromatic acids from wastewaters using methyl isobutyl ketone as solvent. Numerous other applications are under investigation in connection with treating petrochemical and chemical process wastewaters.

Activated carbon adsorption

Activated carbon adsorption has been developed in recent years as a method for recovery of certain types of water-soluble organic chemicals from wastewater and byproduct streams. Although finely divided activated carbon can be used in batched processes, continuous operation using granular activated carbon in a column system is generally much preferred for such applications. A more efficient use is made of the adsorptive capacity of the carbon while columns allow continuous operation and can cope with wide fluctuations in wastewater flow rate and strength. In addition, granular carbons are available which can be reactivated thermally for reuse, minimizing consumption of expensive carbon and eliminating waste disposal problems associated with pulverized carbon.

To determine whether and to what extent the solute can be removed from the wastewater stream by activated carbon, the adsorption curve can be determined in the laboratory. Distribution of the solute between the wastewater and the activated carbon surface at equilibrium is given by the Freundlich equation:

$$x/m = k \cdot C^{1/n}$$

where x = quantity of solute adsorbed
m = quantity of activated carbon
x/m = concentration of solute on activated carbon
C = concentration of solute in solution
k and n are experimentally established parameters.

The adsorption curve is determined by adding different amounts of activated carbon to fixed volumes of samples of known concentration. The concentration of solute still in solution is determined after equilibrium is established.

From the adsorption curve the activated carbon consumption of the waste to be purified can be calculated as:

$$m = \frac{C_i - C_e}{(x/m)\ C_o}$$

where C_i = initial concentration of solute in the wastewater
C_e = desired concentration of the solute remaining in the wastewater after treatment
$(x/m)\ C_o$ = maximum adsorption capacity
m = quantity of carbon necessary to reduce C_i to C_e.

From the results it is possible to determine under static conditions at equilibrium the quantity of carbon necessary to purify a certain wastewater to a desired level. In addition, the most effective type of carbon can be selected for the removal of a particular solute and other factors can be investigated such as the effects of pH adjustment and temperature.

It should, however, be remembered that performance of an activated carbon column under dynamic conditions cannot be fully predicted from laboratory tests carried out under static conditions. Pilot plant testing of the actual wastewater stream under the fluctuating conditions of flow rate, concentration of solute and other parameters to be expected in normal plant operation is necessary. Companies operating in the field of activated carbon treatment of wastewaters are often able to provide a pilot plant facility for testing the effectiveness of solute removal under dynamic loading conditions.

The extent of adsorption of organic compounds to the surface of activated carbon depends essentially on molecular structure, molecular weight and the polarity of the molecules. Adsorption is favoured by planarity and low polarity. Organic compounds with less than four carbon atoms in their carbon skeleton will not be adsorbed. Aromatic compounds are often strongly adsorbed onto activated carbon columns. Phenolic compounds will be adsorbed from acid solutions, where the free phenol is present, and desorbed back into alkaline solutions as the phenate anion. This procedure constitutes an effective method for the recovery of many phenolic compounds from process wastewater streams. Higher-molecular-weight organic acids can similarly be concentrated onto activated carbon from low pH solutions, and subsequently stripped off using high pH solutions of sodium hydroxide. Recovered compounds can also be removed from activated carbon columns by solvent washing or steam distillation.

Another use for activated carbon columns is to remove toxic and inhibiting constituents from process wastewater streams prior to biological treatment of wastewaters. Many of the compounds which give rise to difficulties in biological treatment, including chlorinated aromatic compounds, certain phenols (such as nitrophenols), surfactants and detergents, are readily removed by activated carbon columns. Thus, segregation and carbon treatment of wastewaters containing harmful constituents, before they are mixed with those which contain no compounds harmful to biological treatment, can greatly reduce problems in biological treatment of petrochemical wastewaters. It should be noted, however, that activated carbon treatment is rarely a straight alternative to biological treatment for wastewaters which can be treated by either method, since biological treatment is much cheaper on a BOD/COD removal basis.

Usually several activated carbon columns are required, so that as one becomes exhausted the wastewater can be switched to a fresh column. Often two or more activated carbon columns are used in series to ensure that no harmful compound is allowed to escape into biological treatment. Granular activated carbon columns can be readily blocked by suspended solids when operated under conventional gravity outflow conditions. A limit of 25 mg l^{-1} of suspended solids is usually set for the wastewater and in most cases this means that prior filtration is required. Rapid gravity sand filters are often used for this purpose, where backwashing to remove the suspended solids filtered out is quite convenient. Alternatively, activated carbon columns can be operated in an upward flow mode, where the bed is expanded and suspended solids can pass through the column. The efficiency of the granulated activated carbon to remove the dissolved solute should not be greatly impaired.

A particularly important limitation in connection with petrochemical works wastewaters is the need to restrict the concentration of free oil in the incoming wastewater stream. Normally a limit of $5\,\mathrm{mg\,l^{-1}}$ is set on the free oil concentration, above which the efficiency of the process will be impaired.

Under certain conditions granular activated carbon beds can give rise to the production of hydrogen sulphide, creating odours and corrosion problems. Chlorine or hypochlorite can be dosed into the column to control biological growth.

Provided that due care is taken in the operation of activated carbon columns they can be very effective in removing toxic and non-biodegradable compounds. Operating costs can be high because of the need to replace the activated carbon quite frequently when treating highly polluted wastewaters. The activated carbon can often be thermally regenerated, but the thermal regeneration plants are costly and complex to operate in a non-polluting manner. It is rarely economic for plants utilizing activated carbon to install their own thermal regeneration units; this is normally carried out by the contractor who supplies the activated carbon columns initially.

The thermal regeneration process involves heating the carbon to around $925°-980°C$ in an atmosphere low in oxygen. At this temperature the organic pollutants adsorbed on the surface of the carbon are volatilized or oxidized and a fresh surface of activated carbon is generated. Usually, a multiple-hearth furnace is employed with an afterburner on the top hearth to combust the offgases. Complete oxidation of these offgases is essential to remove the possibility of air pollution by highly toxic substances. The carbon loading is some $215-290\,\mathrm{kg\,d^{-1}}$ per m^2 of hearth surface area and fuel is used to the calorific value of some $18\,600\,\mathrm{kJ}$ per kg of carbon reactivated. A good deal of ancillary equipment is required and close process control is necessary to ensure that the minimum amount of carbon is lost in the regeneration process.

Plants are typically designed to lose up to 10% of the spent carbon during the reactivation process. If the activated carbon is being taken from the industrial process site to a central thermal regeneration facility, transport costs must be added into the overall wastewater treatment costs. Furthermore, activated carbon in wet, static situations is corrosive toward mild steel because of the chemical activity of carbon, so that the columns and any piece of plant in contact with activated carbon under static conditions must be lined with an inert polymeric material, or made from stainless steel.

3.6.3 Treatment to remove oil

An important problem with petrochemical works wastewaters is the quantity of immiscible oil which they may contain. As discussed above, significant quantities of immiscible oil interfere with the biological treatment processes that are necessary to deal with the dissolved organic material and its significant BOD, where wastewaters are to be discharged to inland surface waters. Even where petrochemical works wastewaters are to be discharged into the sea, it is necessary to remove oil to very low levels.

The immiscible oil in wastewaters can be present in a number of physical forms, such as free floating oil, small oil droplets throughout the wastewater, oil attached to solid particles in suspension, and emulsified oil. Every effort should be made to keep to a minimum the amount of oil in small droplet form or in emulsion. This means that turbulence should be minimized, and wastewaters allowed to flow by gravity wherever possible, rather than by pumping. Emulsified oils may require the use of strong chemicals such as acids to break the emulsions, and these chemicals then add to the complexity of wastewater treatment.

Gravity separators

Oily wastewaters are almost always first treated by passing through a gravity separator. Traditionally, petrochemical works employ rectangular gravity separators which are based on the recommended design of the American Petroleum Institute (API, 1969). Alternatively, circular clarifiers have been employed, and more recent years have seen the application of packed plate separators to this task.

The API separator is a large rectangular tank, designed to slow down the wastewater flow rate so that oil droplets can rise to the surface as the wastewater passes slowly along the length of the tank (Fig. 3.3). The key physical parameters of droplet behaviour are the difference in density between the droplets and the water, the size of droplets and the viscosity of the water (a function of temperature). For effective separation, the rate at which oil droplets rise to the surface and can be skimmed off must be greater than the linear rate at which wastewater flows out of the outlet end of the tank. This is equal to the inlet volumetric flow rate divided by the surface area of the tank. Thus, for fixed

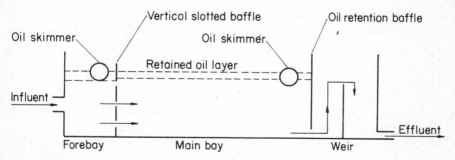

Fig. 3.3 API separator — longitudinal section

droplet parameters, the effectiveness of an API separator will depend in the first instance on its surface area size, while the depth of the tank and its volumetric retention time are of relatively minor importance. The art of designing API separators therefore lies in minimizing turbulence in the tank which could give rise to downward movement of small oil droplets. The incoming wastewater stream first passes through coarse screens to an inlet bay, separated from the main bay by a set of vertical concrete pillars, which are quite close together to avoid wastewater passing into the main bay without being slowed down. This initial slowing of the wastewater flow rate allows the larger oil droplets to rise to the surface of the inlet bay, where they can be removed by skimmers.

The flow containing smaller oil droplets and fine suspended solids is then allowed to pass under an oil-retaining baffle and into the main bay of the separator. Here small oil droplets can rise slowly to the surface, while suspended solids settle to the bottom of the tank. The size of oil droplets removed is a function of the surface area of the bay which in turns depends on costs. Normally, separators are designed to separate down to $150\,\mu$m. Solids which are coated with oil, or oil droplets which have attached themselves to solids, may neither sink nor float, and significant quantities of oil may pass through the separator in this way.

The oil which floats to the surface of the main bay is again held back by a baffle. The purified wastewater is constrained to pass under this baffle and over a weir before it can pass out of the separator. The floating oil is skimmed from the surface of the main bay, avoiding turbulence as much as possible. Rotating disc skimmers are recommended as these give minimum turbulence and recover quite dry oil under continual operation. The type and quantity of solids which collect in an API separator will vary according to the nature of the refinery operation. Where a large quantity of coarse solids is expected to collect it may be necessary to install a sludge scraper so that sludge solids can be scraped to a tank bottom hopper and removed. Normally, in a petrochemical plant, a relatively small amount of fine sludge is collected, which can be pumped from the bottom of the tank intermittently. Any heavy solids which collect can be removed when the separator is drained every few years for maintenance and overhaul.

Usually, in a large works, a considerable number of rectangular API separators are placed side by side so that flexibility can be achieved in terms of standby capacity, faults and routine maintenance. A disadvantage of circular separators or clarifiers is the loss of space between tanks. In addition, it has proved difficult to achieve an even distribution of water flow and avoid turbulence with this design of tank.

The great disadvantage of conventional API separators is the large size of tanks required to achieve good separation, with high capital costs and land requirements. Since the key parameter is surface area, alternative methods for achieving more effective surface area in less space have been sought. An important development has been the parallel plate separator. This has a

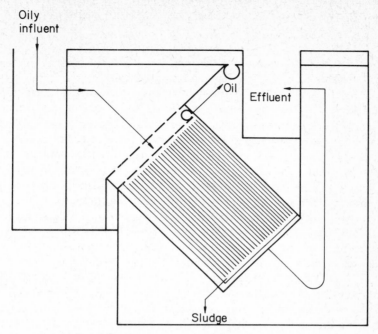

Fig. 3.4 Tilted parallel-plate separator for oil removal

number of parallel plastic plates stacked above each other, with a spacing of a few centimetres (Fig. 3.4). The oil droplets in each space rise until they reach the plate above, where they collect on the underside. Any solids present can settle onto the top surface of the plate below. If the plates are tilted the collected oil can run to the top of the separator, while the sludge can slide down each plate and be collected at the bottom. Each plate in theory adds its whole surface area to the overall effectiveness of the separator. Thus, parallel plate separators can be very compact compared with API separators for the same efficiency of treatment. They can be operated full of liquid, thus avoiding problems of vaporization.

The main disadvantages of parallel plate separators are that they can be readily blocked by viscous oils or sludges and have little capacity to retain large volumes of oil inadvertently discharged to the wastewater treatment plant. They may also encourage the growth of biological films on the surface of the plates. For these reasons such separators are less robust than API separators and require more attention, particularly regular cleaning. They will find increasing application as a second-stage treatment after API separators since they can remove oil droplets down to 60 μm diameter. A recently announced application is the use of these corrugated plate interceptors (made of glass-fibre reinforced plastic) to separate oil from rainwater at a refinery in Essex. This plant is installed in concrete pits some 6 m deep and supported on concrete piles. The plant will treat oily rainwater to a standard whereby it can be discharged into sewers.

Flocculation and sedimentation or flotation

The wastewater from an API separator may typically have a free oil content ranging from 10 to 100 mg l^{-1} of oil with a suspended solids concentration of 20–100 mg l^{-1} and a BOD of 100–500 mg l^{-1}. In general, this wastewater requires further treatment to reduce the residual oil concentration before it can be discharged to sea or passed on to biological treatment. The problem with residual oil lies largely with the small particle size of droplets, although oil solids and emulsified oil may also be present. A number of alternative methods have been developed for removing most of this residual oil, including flocculation and sedimentation or flotation, use of coalescers, and filtration.

The traditional process is flocculation and sedimentation (Oldham, 1979). In this process the wastewater is mixed with a solution of an aluminium or ferric salt and the pH is adjusted until the hydroxide of the metal ion precipitates, pulling residual oil out of the wastewater at the same time. If aluminium sulphate or chloride is used the pH is adjusted to about 6–6.5; if ferric sulphate or chloride, the pH is adjusted to about 8. The hydroxide is then allowed to form into flocs using slow stirring or agitation: fine oil droplets, suspended solids and even some emulsified oil will be absorbed onto the flocs. Flocculation is then followed by sedimentation in a circular clarifier.

Alternatively, dissolved air flotation may be utilized to separate the flocs. In some cases the system may be treated with a polyelectrolyte to enhance floc formation. Recovery of oil for reuse is more difficult with flocculation systems than other methods, since the suspended solids and other flocculated material will be removed with the oily flocs.

In the past, oil recovery involved washing the sludges with dilute acid to dissolve the metal hydroxides and liberate free-floating oil. However, this procedure leaves a highly polluting acid waste which has to be disposed of, making the overall process rather uneconomic. More commonly, the whole mass is thickened in a thickener tank, dewatered by centrifuging or vacuum filtration and incinerated. The other main drawback of flocculation is the continual use of chemicals and the need for close control of plants.

Air flotation techniques have been widely employed in treating oily wastewaters, usually involving the use of polyelectrolytes (or flocculating agents as described above) to achieve coalescence. The effectiveness of any form of air flotation depends on the extent to which the small air bubbles produced will attach themselves to oil droplets and suspended solids, thus pulling them to the surface of the unit. Dissolved air flotation has been used successfully for some types of wastewaters, with the variation of using an inert gas such as nitrogen when air might give rise to a fire or explosion risk. Indeed, air flotation in which under-water rotors with hollow vertical shafts suck air into the water and disperse it has also been utilized. The suspended matter coalesces, rises to the surface and is skimmed off by means of slowly rotating paddles. The operating cost of any such unit is high, owing to the cost of polyelectrolytes, and the performance is variable.

Coalescence
Coalescence de-oilers have been developed in recent years in an attempt to get away from wastewater treatment systems which require chemicals to be added. A number of systems have been marketed, including units incorporating fibrous materials with specific structural oil-release and surface properties and units containing media particles dosed with chemicals and surfactants to make them oleophilic.

One fibre system employs fine alumina fibres to discharge the negative zeta potential on small oil droplets. In this way droplets collect on the fibres and coalesce. For effective operation of such a coalescer, the release of the coalesced oil droplets from the exit phase must be efficient, with minimum redispersion or breakdown of the large oil droplets into smaller ones. Synthetic fibres have been developed which are extremely oleophilic and hydrophobic and can be purpose designed for specific oil-release properties. It has been claimed that wastewater from an API separator containing $50\,mg\,l^{-1}$ of oil with an average droplet size of $20\,\mu m$ can be treated by this method to produce a final oil content of $10\,mg\,l^{-1}$ with an average droplet size of $10\,\mu m$.

Unfortunately, coalescers based on fibre media tend to filter out suspended solids, which are a significant part of the overall problem with most oily wastewaters. Since such units cannot be effectively backwashed, the pressure across the unit can quickly build up to the point where treatment cannot continue and the fibre media cartridges have to be replaced. Effort is now being put into the development of coalescers using coarse granular material which does not get blocked by suspended solids. Such units require preconditioning as a coalescence aid. Preconditioning may require one or more of the following chemicals:

- polyelectrolyte to work as a coagulant;
- surfactant to make the coalescing media oleophilic;
- fuel oil to enrich the media.

These chemicals are dosed to the oily water inlet of the coalescer, the first two usually in very low quantities (about $1\,mg\,l^{-1}$), if at all. The coalescer works by a physicochemical action of drawing tiny oil droplets together, separation of discrete larger droplets within the coalescing media and production of large oil droplets which collect at the top of the vessel.

The plant (Fig. 3.5), which normally operates in an upflow mode, consists of:

- a maturation zone ahead of which an organic polymer and/or surfactant can be dosed;
- a coalescence zone consisting of an oleophilic granular bed 1.5 m deep;

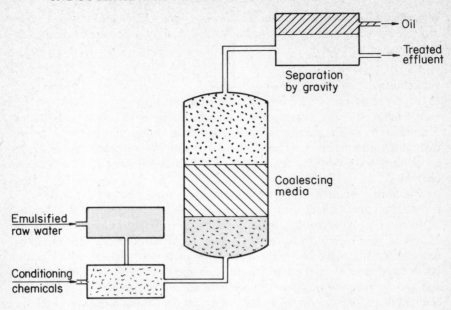

Fig. 3.5 Coalescence de-oiler (Degremont type)

- a separation zone with extraction systems for both oil and treated wastewater;
- a system of air and water backwashing.

Although the medium is progressively clogged by filtered-out suspended solids, it can be readily backwashed using air and water flushing to return it to a clean condition. The plant is said to be capable of reducing the oil content of oily wastewaters from several thousand mg l^{-1} to 10–15 mg l^{-1}. The preconditioning action also allows effective treatment of emulsions other than stable chemical emulsions. The recovered oil (extractable to less than 20% water) can be reintroduced into process operations.

Filtration

Perhaps the simplest and most satisfactory solution to the problem of residual oil removal lies with fine granular filters. While these are commonly referred to as sand filters, they may employ a variety of filtering media including silica or garnet sand, crushed anthracite or various combinations of these in layered beds (Oldham, 1979). These filters can be backwashed with water and air and the backwash water returned to an API separator. If the backwash water is mixed with the other incoming oily water to the API separator there is unlikely to be a separation problem. To allow continuous operation, a number of filters are employed in parallel, so that one can be backwashed while the rest are operational. It is claimed that virtually all the suspended oil present in wastewaters after API separator treatment can be removed by the use of such simple media filters. The obvious limitation of such systems is their inability to treat fully emulsified oils, such as those formed in the presence of detergents. Again, polyelectrolyte dosing may overcome this problem in many cases.

Fully emulsified oils can be broken down by the use of chemicals such as salt solutions or acids. Alternatively, ultra-filtration can be employed. In this technique, the wastewater is passed into a unit consisting of semi-permeable membranes through which the wastewater can be filtered under pressure. The high-molecular-weight oil is not able to pass through the filter medium, but surface-active agents of small size can do so. The emulsion therefore breaks down and oil droplets begin to form. Unfortunately, any suspended solids present will also fail to pass through the membrane and hence give rise to clogging. The semi-permeable membranes are also sensitive to certain chemicals and high temperatures, so are not usually suitable for oil removal from final mixed petrochemical wastewaters. Ultra-filtration may become more widely used for removing oil from otherwise 'clean' wastewaters as it has the advantage of not requiring addition of chemicals, and produces oil with a relatively low concentration of water.

3.7 Biological treatment processes

3.7.1 Pretreatment
As discussed earlier, petrochemical plant wastewaters are very varied, and may contain constituents that are harmful to established processes of biological treatment. While small quantities of biocidal substances can be accommodated by certain biological processes, it may be essential to ensure that certain wastewater streams are not directed to the biological treatment plant. Such wastewaters can be treated using physicochemical processes such as those described, or transported away as wastes for treatment elsewhere. Oil or other immiscible liquids may be present in wastewaters and can be removed by the processes described above.

Many biological treatment processes can adapt themselves to accommodate chemicals which have biocidal characteristics and which would harm conventional sewage works processes if released as intermittent discharges. Thus, biological treatment plants have been developed for the treatment of wastewaters containing quite high levels of simple phenols and aldehydes, where micro-organisms which are particularly able to break down such compounds flourish (Singleton, 1976). The process of adaptation may take some time, but the applicability of biological treatment methods to the purification of certain wastewaters should not be discounted on a superficial examination of biological properties of key constituents.

Often, the only way to determine whether a wastewater will be amenable to biological treatment, or whether one process will be more effective than another, is to carry out laboratory-scale and then pilot plant tests. Small units in which the activated sludge process can be carried out on the laboratory scale are available. Similarly, some indication of whether a biological film will grow on a fixed film aerator can be determined by using a laboratory-scale rotating disc treatment unit. Even more simply, polypropylene tubes have been rotated at an angle to the vertical, and test wastewater passed through them, to determine the rate of growth (if any) of a biological film on the tube walls. An important aspect of test programmes is the examination of the need for nutrient addition. This is likely to be required for petrochemical wastewaters.

The use of large balancing tanks is strongly recommended prior to biological treatment of petrochemical works wastewaters. These equalize fluctuating flow rates of wastewaters from different plants, BOD loadings, shock loads of toxic substances, and pH. In many cases the pH can be balanced so that no specific pH control is required. This is particularly the case in modern works where chemical treatment processes have been replaced by thermal catalytic processes. If pH adjustment is required this should be carried out using automated control systems activated by pH electrodes. Alkaline wastewaters are normally neutralized by either hydrochloric or sulphuric acid, although certain plants may have carbon dioxide available from the flue gases, or even sulphur dioxide or nitrogen dioxide. The choice of a basic reagent for neutralizing acidic wastewaters is usually between sodium hydroxide and various limes, although ammonia or ammonium hydroxide might be available. Despite their higher costs, hydrochloric acid and sodium hydroxide solution may be particularly attractive for neutralization processes since the resulting soluble products do not require a solid–liquid separation step. These compounds are also more reactive than alternative acids or alkalis. Where total dissolved solid levels are causing concern there may be an advantage in precipitating out a neutral salt.

3.7.2 Biological treatment of low- and moderate-strength wastewaters

Many petrochemical works wastewaters, after treatment of process streams to remove as much organic loading at source as possible, have only moderate or low BOD–COD strength when they arrive at the final wastewater treatment plant. This will depend also on the standard of general housekeeping and the nature of the petrochemical processes employed at the works. Wastewaters that have a BOD–COD strength of the same general magnitude as domestic sewage can in principle be treated by biological processes similar to those used for sewage, provided that levels of toxic and non-biodegradable chemicals are low, and nearly all the immiscible oil and solvents are removed before biological treatment. Low quantities of harmful constituents such as phenols or sulphides can be broken down in aerobic biological treatment processes to relatively harmless substances. Thus, sulphide may be oxidized to sulphate and phenol ultimately to carbon dioxide.

The choice of biological treatment method for such wastewaters will usually lie between percolating filtration, activated sludge treatment and, where land is readily available, aerated lagoons. Rotating disc treatment units are also worthy of consideration, and variations of the conventional oxidation ditch may also merit attention.

A range of methods introduced in recent years for the biological treatment of very strong organic wastewaters is considered later in connection with strong petrochemical works wastewaters.

Percolating filtration

The principles and methods of constructing and operating conventional percolating filters are well established (Klein, 1966). Conventional percolating filters employing mineral media have been effectively employed to treat petrochemical works wastewaters for many years. They are extremely robust in normal operation, requiring little operator attention or maintenance. Although capital costs are high, operating costs including energy costs are low, and their relatively high requirement for space can usually be accommodated within the large area of a modern petrochemical works.

Percolating filters are resistant to shock loads in terms of flow rates, pH fluctuations, temperature changes and even, to some extent, toxic chemicals. Recovery rates from severe biological damage may be very slow in cold weather, however. Oil present in wastewaters passed to percolating filters is harmful. The smothering effect of the oil reduces the efficiency of the biological film, while the oil may adhere to the film to such an extent that blockage or 'ponding' of the filter may result. Where percolating filters are to be used it is essential that a reliable and effective method of removing suspended oil is employed beforehand.

Under good operating conditions, the conventional percolating filter, loaded to normal design limits, will readily achieve a 20:30 standard for BOD and suspended solids. Indeed, the humus solids from the filter may be sufficiently low that the normal humus removal stage can be omitted and replaced by a small lagoon (Oldham, 1979).

A relatively recent development has been the introduction of loose plastic media in place of conventional granular media for treating petrochemical works wastewaters. Plastic media, usually made of polypropylene or PVC, have the advantage that they are very light and have a much greater ratio of surface area to volume, with larger voids for air and wastewater to circulate. Thus, the filters can be smaller for the same amount of treatment. They can be placed on lighter foundations and built much higher on a smaller diameter. Higher flow rates can be employed where this is advantageous. Overall, plastic media have greatly reduced the area required for percolating filtration, and although the packings may be more expensive per m^3 than mineral media, this is outweighed by their great efficiency and lower installation costs.

Rotating disc treatment units

These consist of spaced vertical circular discs (or sometimes cages of random fill media) which are mounted on a horizontal axle and rotated partially submerged in the wastewater to be treated. Discs are made of various plastics and in a variety of geometrical forms. They vary in size with the size of the plant but can be 1–4 m in diameter or even larger. They are often mounted in stages, with up to 50 discs in each stage, and perhaps four or five stages in series. As the discs are rotated, micro-organisms present in the wastewater adhere to the disc surface where they multiply. By the end of one week from start-up the disc areas are normally covered by the biomass, although maturity may not be reached for several weeks.

The discs are rotated at speeds varying from one to five revolutions per minute, although the peripheral speed (up to $18\,\text{m min}^{-1}$) is more important than rate of revolution as such. The biological mass in the system, inclusive of suspended material and the biomass of the discs, may be in excess of $18\,000\,\text{mg l}^{-1}$. This is far in excess of that achievable in a conventional activated sludge plant. Daily BOD loading may vary from 10 to $20\,\text{g m}^{-2}$ of disc area with a retention period of about six hours. Some 0.5–0.6 kg of dry sludge solids are produced per kg of BOD removed at about 80% organic content.

Although the capital cost of rotating disc units is inevitably high, the operating costs are low as these units are relatively robust and are certainly low in energy consumption (and noise level) compared with activated sludge units. They have advantages over percolating filters in that:

- they are not visually intrusive (indeed, the units can be sunk to ground level);
- they can be covered, thus removing smell and fly nuisance;
- they can readily be protected against cold weather, which reduces the efficiency of percolating filters.

They are particularly suitable for wastewaters whose flow rate fluctuates widely or ceases overnight.

Activated sludge units

Again, the activated sludge process as conventionally operated is so well established that there is no need to outline its operation here (Klein, 1966). Although the process has been widely used for treating petrochemical works wastewaters, there are now recognized to be a number of significant problems which reduce its attractiveness, particularly for the relatively low BOD wastewaters produced by some petrochemical works. The early attraction of the process for use with petrochemical works wastewaters was based to some extent on the ability to deal with the presence of oil in the wastewater to be treated in a way that could not be achieved with percolating filtration. Commonly, iron salts are added to the influent wastewater together with sufficient alkali to precipitate a floc of ferric hydroxide. As described above, this floc absorbs the residual oil while providing a convenient medium for the activated sludge itself to grow on.

It is well known that the activated sludge process is sensitive to shock loads of toxic substances, temperature fluctuation, pH variation and so on. All of these are liable to arise commonly with petrochemical process wastewaters. In addition to immediate loss of efficiency, bacteria may be rendered inactive for some considerable time, and bulking of the activated sludge in the secondary settlement is another possible difficulty. A less obvious problem is the possibility of failure to produce and separate sufficient activated sludge for return to the system. This can occur if the influent BOD level is rather low and the flow rate exceeds the design flow rate of the secondary settlement tank. Experience has shown that when the influent BOD is below about $100\,\mathrm{mg\,l^{-1}}$ or the temperature of the wastewater can rise above about 40°C, the activated sludge process is not suitable. In such circumstances, provided that low oil concentrations in the wastewater can be assured, it is usually preferable to employ fixed film processes or possibly aerated lagoons.

The more sophisticated variations of the oxidation ditch (extended aeration activated sludge) now being reported for treating a wide variety of organic wastewaters from manufacturing industry may well have a role to play in the treatment of wastewaters from petrochemical works.

Aerated lagoons

The value of aerated lagoons and other forms of relatively simple biological treatment has been somewhat overlooked in European countries. While it is true that the land requirement is high, and they tend to work best in warm, sunny climates, they have a good deal to commend them for treatment of low BOD wastewaters. Petrochemical works are often located in coastal areas or on estuaries where low-cost land is readily available.

If a wastewater of BOD between 50 and $100\,\mathrm{mg\,l^{-1}}$ is passed through a large, shallow lagoon with a retention time of the order of several weeks or more, natural biodegration will take place and discharge consent limits can be achieved. Often, a more effective treatment can be achieved by the use of floating mechanical aerators to increase the oxygen content of the water. Even with such mechanical aids, lagooning is an extremely cheap wastewater treatment method once land costs are discounted. With petrochemical works it is important to ensure that all oil and oily sludge is removed prior to lagoon treatment to avoid the build-up of noxious matter which would eventually have to be removed at great cost.

Final wastewater treatment

Provided that preliminary and biological treatment followed by settlement is satisfactorily carried out there may be no need for further treatment of petrochemical works wastewaters. However, where final wastewater discharge is to an environmentally sensitive watercourse, further treatment may be necessary. Normally, final wastewater 'polishing' involves further removal of suspended solids to achieve still lower BOD and suspended solids levels, commonly by some filtration method.

However, for petrochemical works wastewaters the problem is more likely to lie with residual COD, arising from dissolved organic compounds which resist biological treatment. Indeed, there may be defined toxic compounds present which would be potentially harmful in the water course. In such cases it may be necessary to consider the use of activated carbon treatment as a tertiary treatment method. An alternative approach which has been introduced in North America is to add powdered activated carbon to the aeration basin in the activated sludge process. This possibility will be discussed later in the context of treating strong wastewaters containing organic compounds of low biodegradability.

3.7.3 Biological treatment of high-strength wastewaters

For low- and moderate-strength BOD wastewaters, the biological treatment processes discussed above may be quite satisfactory in terms of BOD reduction. Indeed, the conventional activated sludge process and percolating filtration using mineral media have been widely used to treat petrochemical and chemical plant wastewaters. However, certain petrochemical plant wastewaters can contain high levels of biodegradable organic compounds and have BOD concentrations in the range $15\,000 - 25\,000\,\mathrm{mg\,l^{-1}}$ or higher. Admittedly, the overall BOD of a final wastewater will generally be lower than this, once high BOD wastewater streams have been mixed with lower-strength wastewaters. However, in general, petrochemical plant wastewaters are likely to have organic loadings which are much greater than those of oil refineries alone. This problem of petrochemical works wastewaters is being exacerbated today by the essential water conservation programmes which have been quite rightly introduced.

Several potential problems arise when considering the treatment of high BOD wastewaters by conventional processes. Some processes may simply be unable to accommodate the strength of the wastewater. This often applies with conventional percolating filters where overloading by high BOD wastewater gives rise to excessive growth of biological film and blockage of the filter to hydraulic flow. Although conventional activated sludge plants can deal with much stronger BOD wastewaters than was once considered possible, if designed for the purpose (or updated), they are unable to cope with the shock loads of very high BOD wastewaters which can arise in petrochemical plant operations. The cost of installing and using conventional treatment processes may be very high because of the sheer amount of plant required to give sufficient capacity to treat very strong wastewaters. The land area required for installation of very extensive conventional wastewater treatment plants may not be available.

In view of these problems, it is fortunate that a range of biological treatment processes for dealing with strong organic wastewaters have been developed in recent years. While some of these have been developed with a particular view to the problems of treating strong wastewaters in the chemical and petrochemical industries, others have been developed and exploited initially for use in other industries. However, even processes developed for the food and drink industries are proving to be of great value for treating strong wastewaters from the petrochemical industry.

The major process developments are:

- the use of high-rate percolating filters based on plastic media;
- the use of oxygen instead of air in the activated sludge process;
- the 'Deep Shaft' high pressure variation of the activated sludge process;
- anaerobic biological treatment.

The first two of these processes are now well established, but require some further discussion in the context of treating strong chemical wastewaters. The 'Deep Shaft' process was initially developed to treat very strong chemical wastewaters and, while promising, is still at the stage of being modified and refined on the basis of operating experience. Anaerobic biological treatment processes have been known about for many years, but have only recently come to the fore in connection with treatment of strong industrial wastewaters. There are reasons to believe that they may soon become methods of major importance for treating strong industrial wastewaters. Again, the likely need for nutrient addition to petrochemical wastewaters to achieve efficient biological treatment is stressed.

High-rate percolating filtration

Although there is no strict cut-off point between low-rate and high-rate percolating filtration, the process is usually termed 'high-rate' when the daily BOD loading exceeds 2 kg per m^3 of medium. The most suitable media for high-rate percolating filtration of industrial effluents are considered to be plastics, either in the form of specially fabricated parallel plastic sheet blocks, which can be built up like large bricks to form a tower, or a random-fill small plastic medium of nominal size >50 mm. A variety of plastic media for either application are now available and no one geometry or type of plastic yet dominates the market.

Both polystyrene and polyvinyl chloride (PVC) are used for plastic filter media. PVC tends to be the better material in that it is more resistant to chemical attack, has low flammability and does not become brittle with age. Polystyrene is attacked, to some extent, by various organic solvents and may be unsuitable for use with effluents containing such solvents (as is likely for petrochemical plant effluents).

The design of medium for efficient high-rate percolating filtration must be a compromise between provision of the maximum area of surface to support the microbial film, and the maintenance of voids which will not become choked and will allow the wastewater to be evenly distributed, and also allow access by air to all of the biological film.

The amount of film which develops on the surface of the medium depends on the BOD loading, which in turn is a function of the BOD of the effluent and the hydraulic loading. Normally the surface hydraulic loading (about 15 m^3 m^{-2} d^{-1}) should not be greater than that necessary to ensure complete wetting and therefore maximum use of all the medium, so that the retention of the liquid in the filter is maximized. To attain this result efficient distribution of the wastewater over the surface of the filter is essential. For wastewaters of very high BOD and containing readily biodegradable organic matter, rather higher surface hydraulic loadings (greater than 30 m^3 m^{-2} d^{-1}) may be utilized, using recirculation of effluent to increase the flow rate. This dilutes the incoming effluent, compensates to some extent for non-uniform distribution, and spreads the growth of film more evenly over the entire depth of the filter and over the full surface of the medium.

Usually high-rate percolating filters based on plastic media are first designed and used for about 70% BOD removal. Further reduction in BOD may be achieved by passing the effluent to another high-rate percolating filter or, if the BOD has now been sufficiently reduced, to conventional biological processes. Often a high-rate tower percolating filter based on stacked modules is used to take out some 70% of the BOD strength, with a more conventional percolating filter using random-packed plastic media employed as a second-stage biological treatment to achieve satisfactory effluent discharge standards. Alternatively, the effluent from the high-rate filter may be treated by the activated sludge process or discharged to sewer for municipal sewage treatment, if sufficient BOD reduction has been achieved in the first stage.

High-rate percolating filtration has definite attractions for the treatment of strong petrochemical wastes because of the ability of the process to withstand variation in hydraulic loading and shock loads of high BOD wastewater or toxic constituents. An example of the use of high-rate percolating filtration based on plastic media for the treatment of a petrochemical plant effluent arose from expansion of a Terylene fibre plant on Teesside some years ago (ICI, 1976). This required increased production of terephthalic acid and the construction of a new chemical plant. The high BOD load of the terephthalic acid plant wastewater needed to be reduced to a strength similar to domestic sewage so that it could be mixed with the main plant effluent. The effluent treatment plant was designed to treat 959 m^3 per day of wastewater from two main sources, with a combined concentration of 2500 mg l^{-1} and a load of about 2358 kg BOD per day. The acidic wastes were neutralized by the controlled addition of lime slurry, and the neutralized effluent passed to a primary settling and balancing tank to give about $3\frac{1}{2}$ h retention. The overflow from this tank was pumped to the first of three stages of high-rate percolating filtration using Flocor E plastic medium modular towers and operating in series. The first and second stages were packed with 648 m³ of Flocor E and the third stage with 605 m³ of Flocor E. Nutrients were added to the effluent before the first stage, and recycle was employed, with settlement of humus sludge carried out after each stage of percolating filtration. This plant gave an effluent of BOD strength similar to domestic sewage which could be passed to the main petrochemical plant wastewater stream.

Loose plastic media are tending to replace modular block media in some applications as they have the virtue of flexibility and even higher surface-to-volume ratios. Thus, loose plastic media can be used to replace mineral media which have deteriorated in existing percolating filter plants. The lower profile of loose plastic media plants may be preferable to the higher towers of modular-block plastic media filters. The increase in surface area available may be illustrated by reference to Flocor R random-pack plastic medium with a ratio of surface area to volume of $240\,m^2\,m^{-3}$ compared with Flocor E modular block medium at $82\,m^2\,m^{-3}$.

Uprated aerobic activated sludge processes

Conventional activated sludge processes have been used for treating effluents from chemical and petrochemical plants for many years. Modern designs allow much higher loading rates and MLSS levels, because greater volumes of oxygen in air can be supplied to tanks. Well controlled activated sludge plants can treat quite strong organic effluents and achieve satisfactory BOD reductions. Often, however, the hydraulic loading and BOD strength of wastewaters passed to conventional activated sludge plants has been increased continually over the years so that plants are overloaded. In this situation shock loads of still higher-strength effluents or toxic substances can lead to failure of the system.

One possible method for uprating conventional activated sludge plants is to use pure oxygen instead of atmospheric oxygen as the source of dissolved oxygen in the mixed liquor. Oxygen can now be obtained cheaply enough, using the pressure swing adsorption process, to justify using it in effluent treatment. Pure oxygen can be supplied via diffusers, or by replacing the atmosphere above a mechanically aerated tank. Usually, the tanks are covered to make maximum use of the oxygen.

In diffuser systems, the oxygen escaping through the mixed liquor at the first pass is recaptured and pumped round the system again and again until about 90% of the gas is absorbed in the liquid. In surface aeration plants, the gas space above water level is pressurized with high purity oxygen and the mechanical aerator shafts pass through the roof of the tank to the drive mechanism. Throughout the process the atmosphere of the channels is held at a slight positive pressure.

The exhaust gas consists of about 50% oxygen and 50% other gases (principally carbon dioxide) that have been stripped from the biological process. Because of the continual recirculation of the carbon dioxide resulting from the bacterial breakdown of organic material, the carbon dioxide content of the mixed liquor is appreciably higher than in a conventional process, and consequently the pH of the liquor is lower. Unfortunately, this low pH means that nitrification is not usually possible with the high purity oxygen activated sludge process.

The big advantage of the process is the high loading rate obtainable. This means that existing plants can be uprated, or the tanks installed in new plants can be smaller than would otherwise be required. Any possible smell nuisance is virtually eliminated, and the process has the flexibility associated with modern activated sludge plants. The general disadvantages include the cost of the production of oxygen, the extra complexity of the process and the lack of nitrification. In the context of treating petrochemical plant wastewaters a major disadvantage arises from the hazards associated with the use of pure oxygen. In addition to increasing the overall fire and explosion risk on petrochemical plants, pure oxygen in the presence of any volatile flammable substances in the wastewater will give rise to a serious risk of fire or explosion in the activated sludge plant itself.

An alternative approach to increasing the dissolved oxygen level in the activated sludge process is to operate it above atmospheric pressure. This is the approach employed in the 'Deep Shaft' process developed by ICI. It has been introduced in connection with treatment of very strong organic chemical wastewaters from petrochemical and chemical plant processes (Hemming et al., 1977). The process incorporates a high rate of oxygen transfer with a reduced retention time and a significant reduction in the production of surplus sludge. Basically the system consists of two concentric vertical shafts which are sunk 40–150 m into the ground. The mixed liquor passes downwards through the centre shaft (or downcomer) and rises in the annular space between the cylinders (the riser). Air is injected into the downward-flowing mixed liquor, some distance below the surface. The momentum of the liquor carries the air bubbles down to the bottom of the shaft where the pressure may be between 10 and 15 atms. The oxygen absorption rate is increased by a similar amount compared with the surface rate.

The air is in contact with the absorbing water for a period of up to two minutes before it reaches the deepest point (compared with the 15 s contact time in a conventional diffused air process). It is likely that the rate of oxygen absorption is about ten times that of normal activated sludge processes. As the mixed liquor rises in the outer shaft, the reduction in pressure causes the release of dissolved gases to form bubbles, consisting largely of nitrogen and carbon dioxide since nearly all the oxygen is used up. At the top of the shaft there is a relatively large open disengagement chamber which allows the engaged waste gases to pass out to the atmosphere before the mixed liquor again passes into the downcomer.

A major problem of the 'Deep Shaft' process is the separation of the sludge from the effluent. The sludge, which is buoyed up by entrained gases, is taken to a flotation chamber. Here the sludge can be readily skimmed off and

Table 3.6 Volume and power requirements for 'Deep Shaft' and high-rate activated sludge processes treating biodegradable effluent of 2000 mg l^{-1} BOD to give 200 mg l^{-1} BOD (Hemmings et al., 1977)

BOD load (kg d^{-1})	'Deep Shaft'		High-rate activated sludge	
	Aeration tank (m^3)	Power (kW)	Aeration tank (m^3)	Power (kW)
1800	37.5	26	360	54
6000	125	82	1200	178

returned to treatment, or discarded as surplus activated sludge. Following the flotation chamber the remaining mixed liquor is processed through a degasifier which may be either a reduced pressure unit or a mechanically stirred unit. The remaining solids may then be settled out in the conventional manner.

As a result of the unusual environment, associated with relatively high pressures in which the micro-organisms operate, a higher proportion than usual of the carbon in the waste is emitted as carbon dioxide. This results in an appreciably lower than normal production of surplus sludge. It is claimed that the sludge yield is as little as 0.4 kg per kg of BOD removed, instead of the 0.9 kg to 1.0 kg (dry solids) of digested sludge that would result at the end of conventional treatment processes for every kg of BOD removed. Obvious advantages of the 'Deep Shaft' are the lower aeration volume and power requirements for treating strong effluents, as shown in Table 3.6. This table is based on the treatment of readily biodegradable waste in a wastewater of 2000 mg l^{-1} of BOD being treated to give an exit concentration of 200 mg l^{-1} BOD. As primary settling tanks are not required, and the shaft requires very little space compared with normal aeration tanks, the land space saved is very considerable.

The main disadvantage of the process lies with the construction and maintenance of the shaft itself. The shaft costs per unit volume are appreciably greater than the costs of shallow aeration tanks. As the shaft gets deeper so it becomes increasingly more expensive to excavate each unit volume. Relatively narrow shafts of up to 2 m diameter may be drilled, while larger diameter shafts may be piled or mined. Generally, it is cheaper per unit volume to construct larger diameter shafts. The cost of the shaft is of course related to the geology of the area. Once formed, the shaft may be lined with concrete, plastic or steel, depending on the diameter and on the geological conditions. Because of the construction difficulties and costs, it is unlikely that the 'Deep Shaft' process will be used for other than large sources of relatively high BOD wastewaters.

The other important disadvantage is that nitrification is rarely achievable with the 'Deep Shaft' process, although experiments are in progress to nitrify 'Deep Shaft' treated effluents in a second biological treatment stage.

A number of successful plants are now in operation, both in Europe and in North America, treating strong industrial wastewaters or domestic sewage containing a high proportion of strong organic wastewater from industry.

Generally speaking, it is likely that both the high purity oxygen activated sludge process and the 'Deep Shaft' process will be used to remove the major amount of BOD from strong industrial effluents. A second-stage conventional treatment process can then be used to reduce the BOD level from, say, 300–150 mg l^{-1} to 20–50 mg l^{-1}, to achieve river discharge consent limits, including nitrification of the effluent.

An extension of the activated sludge process which has important applications for treatment of petrochemical wastewaters is the Du Pont PACT Process (Flynn, 1975). In this process powdered activated carbon is added to activated sludge in conventional activated sludge treatment plants. A matrix of micro-organisms and carbon particles is formed. The activated sludge plant is operated in the conventional way with the final settled sludge forming in a dark, well flocculated form. Sludge recycle is carried out as normal with excess activated sludge being formed and removed as before, and some fresh carbon being added to the system.

The activated carbon adsorbs organic chemicals according to their physico-chemical properties, as described earlier, and will adsorb compounds which are not amenable to biodegradation. The carbon also improves overall operation by dampening variations in organic loading, adsorbing toxic or inhibiting substances and surface-active agents that would cause foaming. The presence of activated carbon in surplus activated sludge improves solids handling and dewatering. If carbon recycle is desired, the surplus sludge can be passed to a thermal carbon regeneration process. The dosing of powdered activated carbon can be varied according to the nature of the wastewater to be treated and the desired effluent quality.

Organic compounds that can be treated by this system fall into three categories:

(1) Adsorbable non-biodegradable compounds which are removed only by carbon adsorption.
(2) Non-adsorbable, biodegradable compounds that are removed only by micro-organisms.
(3) Adsorbable, biodegradable compounds that are removed by either carbon adsorption or micro-organisms.

Compounds which are neither adsorbable nor biodegradable will of course not be removed by the PACT system.

An important feature of the PACT process is the regeneration of the activated carbon surface when an adsorbable biodegradable substance is adsorbed onto it. This process seems to take place via reversibility of adsorption of the material, so that as biodegradation of such materials in solution leads to a lowering of their concentration, adsorbed biodegradable materials are desorbed again. In this way sites on the carbon surface occupied by readily biodegradable materials can be made available for less biodegradable materials, even if the former are in much higher initial concentration. The carbon activity will fall off when all available sites are taken by adsorbable non-biodegradable compounds.

Research indicates that microbiological activity does take place on the carbon surface, leading to some biological fouling of the activated carbon surface. One study found that about 30% of the surface of carbon that had been in contact with bacteria in a PACT system for an average of five days was inaccessible for adsorption (Flynn et al., 1976). However, this phenomenon is not strictly a disadvantage since the bacteria seem able to degrade the less biodegradable adsorbable organics when in contact with the carbon for the sludge age rather than the hydraulic retention time. The outcome is that the overall effect of bacteria on the carbon surface is to produce an effluent of improved quality when compared with the additive results obtained by considering biological reaction and adsorption as independent phenomena.

As discussed earlier, pressure is increasing to reduce the concentration of non-biodegradable and slowly biodegradable compounds with possible toxic properties in treated wastewaters from petrochemical works. A great advantage of using the PACT process to achieve this result is that activated sludge plants can be readily converted to include activated carbon dosing. This avoids the need to introduce a completely separate additional treatment process. The addition of activated carbon would seem at first sight to increase the operating costs of the process. However it is claimed that adjustment of carbon dosing introduces extra control into the process, allowing fine tuning to optimize treatment and reduce operating costs. Certainly, by introducing the PACT process more concentrated wastewaters can be treated within existing plant capacity. Du Pont introduced the PACT process at its Chambers Works in New Jersey in 1977 in preference to the use of granular activated carbon columns on the grounds of substantially lower capital investment.

Anaerobic treatment of strong wastewaters

While uprated systems of aerobic treatment of strong organic effluents have been developed in recent years, there has also been strong interest in the possibility of developing anaerobic treatment processes. Certainly, for effluents of greater than about 3000 mg l^{-1} BOD strength conventional aerobic treatment processes are unsatisfactory. It has been suggested that anaerobic treatment becomes economically feasible with wastes of COD strength greater than 4000 mg l^{-1}, and that at COD strength of 20 000 mg l^{-1} the costs of anaerobic treatment will be about one quarter of those of aerobic treatment. Anaerobic treatment offers the possibility of extensive treatment in relatively small plants and the production of low quantities of sludge compared with aerobic processes. The conversion of much of the waste to methane gas can be an additional advantage if the gas can be used as a fuel source on the plant. With rapidly rising energy costs this is becoming an important argument in favour of anaerobic treatment processes.

A full account of anaerobic treatment processes would be beyond the scope of this chapter, and inappropriate in that as yet they have been mainly applied to strong wastewaters from food and drink processing or to wastewaters from intensive livestock-rearing, rather than to petrochemical wastewaters. Anaerobic versions of the activated sludge process are well established, and the last few years have seen the rapid development of anaerobic filters. In anaerobic activated sludge processes, the separation of the sludge after treatment can be problematical since gas bubbles remain in intimate contact with sludge particles. Vacuum degassing can aid settleability, or alternatively dissolved gas flotation can be employed instead of settlement. Effluents from this process, having undergone BOD reduction of 80–95%, are readily amenable to further aerobic treatment if necessary to produce final effluents of river discharge quality. The process is reported as being suitable for industrial effluents with a BOD strength up to $25\,000\,\text{mg}\,\text{l}^{-1}$.

Anaerobic filters are upward-flow systems similar in many respects to aerobic percolating filters, and are usually filled with high-voidage plastic media or a series of plates. There is a definite correlation between organic loading and percentage COD removal, and some filters are able to provide partial treatment (40% COD removal) at daily loadings as great as $8-16\,\text{kg}\,\text{COD}\,\text{m}^{-3}$. However, the effluent is still putrescible and strong smelling. It is probably better to load the process at a lower rate (about $4\,\text{kg}\,\text{COD}\,\text{m}^{-3}\,\text{d}^{-1}$) and to achieve a greater percentage reduction (70–90%). Anaerobic filters apparently withstand shock organic loads well.

Key problems with anaerobic treatment processes lie with the sensitivity of the bacteria, particularly the methane-producing bacteria, to unfavourable environments and inhibiting compounds. The pH of the process is critical; acid produced by the non-methanogenic bacteria requires to be neutralized by ammonium bicarbonate formed by the reaction:

$$CO_2 + H_2O + NH_3 = NH_4HCO_3$$

If the effluent contains little or no nitrogenous matter, there will be no formation of the ammonium bicarbonate buffer — this could well be the case for petrochemical effluents. Addition of sufficient nitrogenous material is then necessary to help pH balance, but care is essential to prevent excessive addition of nitrogenous substances, as the formation of too much ammonia will inhibit the whole process. Methane-producing bacteria are also very sensitive to the inhibiting effects of anionic detergents, chlorinated hydrocarbons and heavy metals.

These problems of sensitivity may hold back the application of anaerobic treatment processes to petrochemical plant wastewaters. However, where inhibiting compounds are very unlikely to find their way into a particular process wastewater that is very strong, anaerobic treatment offers an attractive possibility. When considering the possibility of anaerobic treatment for a particular waste it is essential to be aware that bad odours may arise, particularly if the waste contains sulphur compounds. It is virtually always necessary to follow an anaerobic treatment process by an aerobic treatment process.

A further possibility for the future is the use of particular process wastewaters as a base feedstock for microbial conversion processes leading to valuable products, such as single-cell protein.

3.7.4 Final effluent treatment

The biological treatment processes outlined above give rise to sludges, from which the effluent must be separated. Separation processes are usually based on gravity sedimentation or air flotation techniques, although centrifugation and filtration may be realistic alternatives in some cases. Provided that these processes are carried out satisfactorily, an effluent with a BOD of less than $20\,\mathrm{mg\,l^{-1}}$ and less than $30\,\mathrm{mg\,l^{-1}}$ of suspended solids can be achieved. As discussed earlier, an effluent that contains no toxic or non-biodegradable organic compounds is usually acceptable for discharge to a watercourse which gives adequate dilution.

However, it may be necessary to achieve a much lower level of BOD and suspended solids (such as BOD $5\,\mathrm{mg\,l^{-1}}$; suspended solids $5\,\mathrm{mg\,l^{-1}}$) to meet discharge requirements for a particular river. In such cases, a tertiary or final effluent treatment process may be necessary. Usually this consists of a physical process for removal of some of the remaining suspended solids. Since this material is usually of organic origin, its removal will result in reduction of both suspended solids and BOD levels. Filtration processes are commonly used to remove such suspended solids, including microstrainers and sand filters. Traditionally, sloping grass plots have been used to carry out the filtration. Alternatively, lagoons have been used to hold back the wastewaters for further biological action and sludge settlement. These may not be effective where weather conditions cause turbulence in the lagoon water.

Excessive levels of ammonia are a problem with some wastewaters, particularly where the biological treatment process employed does not favour nitrification. Ideally, the original biological treatment process should be chosen or modified so that account is taken of the need to achieve nitrification of nitrogen-containing wastewaters. Alternatively, a final treatment process to achieve nitrification may be necessary. For petrochemical plant wastewaters this would probably be a nitrifying aerobic filter. Where the original wastewater contains large amounts of nitrogenous material, nitrification may lead to the production of unacceptably high levels of nitrate. In such cases, a biological treatment process to carry out denitrification may be required. Again, denitrification processes based on both the activated sludge system and percolating filtration have been developed in recent years, because of growing concern about increasing nitrate levels in surface waters, especially those used for potable supply.

As discussed above, the presence of traces of non-biodegradable organic compounds in drinking-water, with possible long-term adverse effects on health, is a matter of increasing concern. As a result, there is now much more interest in residual COD levels in treated industrial organic effluents, especially where such compounds as organochlorines are known to be present and where the effluent is discharging into surface water abstracted for potable supply. In such cases, or where there may be adverse effects on fish and other higher organisms in surface water, the removal of some of the residual COD (or specific components) may be required. If the residual COD or undesirable components are in dissolved form, which is most likely, the preferred method of removal will probably be activated carbon treatment. Granular activated carbon columns are already in widespread use to remove traces of higher-molecular-weight organic compounds from certain effluents and to purify water for potable supply or particular industrial uses. As discussed earlier, activated carbon treatment is particularly effective in removal of many of the toxic and persistent organic compounds commonly found in petrochemical wastewaters.

Normally, most of the undesirable compounds are non-polar and can be removed under neutral effluent conditions after biological treatment. Where residual acidic or basic organic compounds are to be removed, pH adjustment may be necessary to achieve satisfactory removal. Coloured wastewaters containing organic dyes and pigments can very often be decolorized by passing through activated carbon. Activated carbon columns have the additional advantage that they can remove trace amounts of suspended solids also, although it is preferable to remove any significant amounts of suspended solids by prior filtration through a suitable sand filter.

3.8 Disposal of concentrated wastes and sludges

Concentrated process wastes, even those containing significant quantities of water, are often best dealt with outside the wastewater treatment system. It is cheaper and more effective to segregate such wastes in tanks and avoid passing them to the treatment plant where they provide shock loads, and may create fire or toxicity hazards. Recovery of valuable constituents from wastes is often possible, especially as the costs of oils and solvents increase. If the petrochemical plant itself cannot incorporate the recovery process, a contractor may be able to accept the material for recovery at his plant, sometimes after bulking with similar wastes. Package solvent recovery plants are becoming a common feature of the industrial scene. For more contaminated hydrocarbon wastes, incineration with heat recovery may be the most economically viable disposal route. Provided that the waste is autothermic, liquid and does not contain too much water, and provided that sulphur-, nitrogen- and halogen-containing substances are largely absent, incineration with heat recovery is usually straightforward.

If the waste contains significant quantities of compounds with bound sulphur, nitrogen or halogens, heat recovery is difficult and the waste should be disposed of in a purpose-designed incinerator with an advanced gas cleaning system. Such incinerators are available and certain ones are designed to cope with viscous semi-solid material, solid lumps and powders, as well as liquids. Some wastes will not sustain combustion and must be sprayed or inserted into a flame or hot gas stream in a special chamber. Alternatively, a fluidized-bed incinerator which accommodates wastes in any physical form may be used. Often, and particularly for organochlorine wastes, the combustion system must achieve a very high temperature to achieve complete combustion. Thus, polychlorinated biphenyls require temperatures well above 1000°C and a significant residence time to achieve full combustion.

The solid residues remaining after incineration may vary in amount and in nature from a solid clinker to a fine powder. It is usually necessary to test such residues to ensure that they do not themselves constitute a toxic hazard due to only partial combustion of the organic material, or contamination by, say, heavy metals.

The flue gases from such incinerators will contain particulate matter and acidic gases from sulphur, nitrogen and halogens if these are present in the waste. The acidic gases, such as hydrochloric acid, are extremely corrosive and it is necessary to build the plant to withstand attack (for example by using stainless steel at key points instead of mild steel). To meet air pollution control regulations, such acidic gases and particulates must be removed from the exhaust gas stream. Particulates are most commonly removed by electrostatic precipitators and acidic gases by scrubbing with water. This water becomes acidic in turn and must be neutralized before discharge to sewer. The exhaust gases must be monitored to ensure that emissions do not contravene regulations.

Incineration is undoubtedly the best method, on pollution control grounds, for the disposal of highly flammable, toxic or very persistent organic wastes. However, the costs of disposal by this method are considerably higher than for land disposal of hazardous wastes. Thus, there is pressure to allow disposal of some wastes of this type on land, and this is sometimes allowed where adequate precautions can be taken to ensure no environmental or occupational health hazard.

Suspended solids or liquid wastes removed in the preliminary stages of effluent treatment may similarly have potential for recovery, particularly after dewatering. Alternatively, they may be disposed of, either by incineration or land disposal according to their character.

Sludges from biological treatment of refinery or petrochemical plants will often be similar in character to other sludges from biological treatment plants. The physical and chemical character of the sludge will determine the most appropriate method of sludge treatment and ultimate disposal. Some sludges may be relatively easy to dewater by filtration or centrifugation, while for others anaerobic digestion may be preferable to render the sludge easier to handle and dewater.

Some sludges may be suitable for disposal by use as a fertilizer on land, while others will be quite unsuitable for this purpose as they contain traces of oil or immiscible solvents, or toxic organic or even inorganic compounds. In this case, disposal into suitably protected landfill sites, or possibly sea disposal, may be appropriate.

3.9 Costs of effluent treatment

A detailed analysis of capital and operating costs for oil refinery and petrochemical plants is beyond the scope of this chapter. The sheer variety of wastewaters, the alternatives for ultimate disposal, wastewater treatment methods and waste disposal routes is bewildering. A thorough study of European petrochemical industry aqueous effluents and their treatment was carried out in 1977 and remains the standard work on this subject (CEFIC Petrochemicals/Ecology Group, 1977). Some of the key findings of this report are the following.

The investment costs for the segregation of sewers are substantial. Investment for sewer segregation and preliminary treatment of effluents prior to biological treatment might add between 20 and 50% to the cost of the biological treatment plant. Indeed, BASF reported an investment for segregation and in-plant improvements exceeding the cost of the biological treatment plant. In general, the running costs of pretreatment plants are, however, low relative to biological treatment plants. Indeed much of the investment and operating costs might be offset in terms of the costs of materials recovered.

The costs of biological treatment for effluents from some twelve plants were considered, taking into account costs of sludge removal or incineration. BOD removal averaged 93%. Land costs were not included. In spite of the variation in biological plant design and the wide range of throughputs and BOD concentrations, mathematical formulae for the investment and operating costs of the plants were derived which fitted most plants' real costs quite accurately. Based on the situation in 1976, and taking the throughput at Q m^3 h^{-1} and the daily BOD at B kg, investment costs were I D.fl (Dutch florins) where

$$\log I = 5.194 + 0.163 \log Q + 0.429 \log B$$

Running costs covering power, chemicals, labour maintenance and overheads were C D.fl per annum where

$$\log C = 4.056 + 0.546 \log Q + 0.218 \log B$$

Fixed running costs (labour, maintenance and overhead) ranged from one-third to two-thirds of total running costs, while variable costs (power, chemicals and services) made up the remainder. However, the split of running costs could not be correlated with the size of treatment plant and probably reflects local circumstances and customs (for example, on manning levels).

The investment costs for biological treatment are very sensitive both to the required average BOD concentration of the final effluent and to the limitations placed on variability of the final effluent. Thus, a long-term average BOD_5 concentration lower than 50 to 30 mg l^{-1} may well require more plant capacity. This may require longer aeration time, second-stage aeration or removal of suspended solids by filtration. If the regulating authorities require the occurrence of excursions to high BOD to be rare, or the ratio of highest BOD in discharges to the average BOD to be low, then special steps which add to costs will often be necessary. These include more flow balancing, increased aeration volume and duplication of equipment in case of failure.

Where mixed media or sand filters are used as a final effluent treatment step to reduce solids and BOD–COD in the effluent before discharge, the investment costs I and running cost C in Dutch florins at 1976 values are given by:

$$I = 0.2 \ Q^{0.66} \times 10^6 \quad \text{D.fl}$$

$$C = 5000 \ Q^{0.65} \quad \text{D.fl per annum}$$

These costs include the cost of designing and operating the overall plant to backwash, separate and dewater the solids removed by the filter columns.

Where activated carbon treatment may in future be required for final treatment of petrochemical effluents, there is no doubt that this will add very substantially to overall capital and running costs. Here the investment costs I and running costs C per annum, again in Dutch florins at 1976 values, will be

$$I = Q^{0.6} \ (0.07 + 0.09 D^{0.6}) \times 10^6 \quad \text{D.fl}$$

where D is the carbon dosage (kg of carbon to be reactivated per m^3 of wastewater). This covers the cost of adsorption and regeneration plant, and

$$C = 200 \ Q + 2500 \ QD + 0.1 \ I \quad \text{D.fl per annum}$$

Here fixed running costs are assumed to be a fraction of the investment costs.

The adsorptive capacity, particularly after several regenerations, has a major impact on investment and operating costs once volumetric throughput and COD reduction are fixed.

The overall conclusions from the study of petrochemical plant effluent treatment costs were particularly interesting. Unless the costs of more advanced effluent treatment plants with full biological treatment (compared with, say, oil separation alone) can be passed on in product costs, the annual costs of central effluent treatment can reduce plant profits by 10–20%. Furthermore, if only a proportion of the plants' capacity can be utilized, which is particularly likely during the present period of recession, then the annual costs of treatment will have a much greater impact, especially since plant profits will inevitably be lower.

Another feature of the overall results is the extra penalty carried by smaller plants treating effluent to the same standard as larger plants. Thus, a model site used in the study as typical of the operations of existing petrochemical plants was found to spend 5.1% of its total capital costs on its effluent treatment plant; annual running costs would be 1.9% of the annual proceeds. For a plant one-tenth of the size of the model site, however, the capital cost of the effluent plant would be 12.8% of the total plant costs, and the annual running costs 3.9% of the annual proceeds. Undoubtedly the future will see an intensified search for cost-effective treatment methods.

3.10 Case studies of wastewater treatment

3.10.1 A resins manufacturing plant (Singleton, 1976)

This plant, established in 1948 and subsequently expanded, is in a rural inland area and manufactures a range of resins. The factory effluent contains methanol, formaldehyde, phenol and small quantities of other compounds, together with traces of resins in solution and suspension (Table 3.7). Each

Table 3.7 Resin factory raw effluent characteristics

Component	Concentration ($mg\,l^{-1}$)
BOD_5	1000–2000
Methanol	400–1000
Formaldehyde	100– 600
Phenol	10– 100

major production plant has its own catchpit which prevents the build-up of resinous matter in the drains. Cooling-water is obtained from boreholes and fed into a water-recirculating system which has enabled water requirements to be reduced from $48\,m^3$ per tonne of resin to $27\,m^3$ per tonne. The discharge from this system, together with other cooling-water from the factory, flows by gravity to the treatment plant where it is used for dilution of final treated effluent. Surface runoff water is collected in stormwater tanks and pumped to the factory waste storage lagoon where it provides additional dilution. Domestic sewage from the factory and from a neighbouring estate of about 180 houses discharges to a sump and is pumped to the wastewater plant for treatment. The dry weather flow is about $200\,m^3$ per day. The sewage is collected in a $9\,m^3$ settlement tank on the effluent plant from which it is pumped to the primary activated sludge plant to aid treatment of the process effluent. The untreated factory wastewater is stored in an earth-embanked lagoon of about $4500\,m^3$ capacity, with a clay floor. This acts as a balancing tank.

Through good housekeeping and process modification to recover materials from waste streams, the BOD strength of the wastewater has been reduced from $4500\,\text{mg}\,\text{l}^{-1}$ in 1955 to something over $1000\,\text{mg}\,\text{l}^{-1}$ in 1975. The rapid expansion of the factory resulted in the wastewater flow rising from about $180\,\text{m}^3$ per day in 1954 to about $1250\,\text{m}^3$ per day in 1975. However, the BOD load has only increased from 1500 kg per day to 1700 kg per day (about 4 kg per tonne of finished product).

As indicated above, the factory wastewater is mixed with domestic sewage and diluted, and can then be treated by the activated sludge process, in spite of the high level of bactericidal organic waste. The biological treatment takes place in two stages, a first stage of activated sludge treatment followed by a second biological treatment stage. Through historical circumstances some of the partially treated effluent is passed through a second activated sludge treatment for secondary biological treatment, while the rest of the effluent is treated by percolating filters. The primary activated sludge process uses mechanical surface aeration with a very high MLSS level of about $12\,000\,\text{mg}\,\text{l}^{-1}$. The dissolved oxygen level in the completely mixed tank system is rarely above $1\,\text{mg}\,\text{l}^{-1}$, but 80–90% of the polluting load is removed. The activated sludge is settled in a conventional circular settlement tank, and the rate of sludge return is about 150% of the influent rate.

The clarified effluent from the settlement tank overflows a peripheral weir and passes into a splitter chamber where a series of penstocks directs the flow to the secondary treatment processes. About 40% of the effluent is directed into a diffused air plug-flow activated sludge plant (which was previously the primary biological treatment plant). The remaining 60% is transferred to two 28 m diameter percolating filters operating in series, each with its own humus tank.

The wastewaters from each of the secondary biological processes are combined and flow through a balancing pond. The effluent from this is mixed at about 1:8 dilution with the surplus cooling-water and discharged to river via a ditch 800 m long. The volume of effluent flowing into the river is about $12\,000\,\text{m}^3$ per day; under very dry weather conditions the river gives as little as 1:1 dilution. The water authority consent conditions for discharge are: BOD and suspended solids $20{:}30\,\text{mg}\,\text{l}^{-1}$, with a limit of $5\,\text{mg}\,\text{l}^{-1}$ for formaldehyde and $1\,\text{mg}\,\text{l}^{-1}$ for phenol. From Table 3.8 it can be seen that these consent limits are readily achieved by the final effluent. Surplus activated sludge and humus

Table 3.8 Resin factory treated effluent characteristics

Component	Concentration $(\text{mg}\,\text{l}^{-1})$
Suspended solids	2.5
BOD_5	2–4
Formaldehyde	0.6
Phenol	0.03
Ammoniacal N	1
Total Oxidized N	10

sludge from the biological treatment plants can be dewatered by centrifuge and are then removed from the site.

Important aspects of this case study include the emphasis on material recovery rather than degradative treatment, and clever use of available resources and conventional technology to deal with a difficult wastewater very cost-effectively. It is estimated that current savings in raw materials account for approximately twice the annual operating costs of the wastewater treatment plant.

The overall capital cost of the treatment plant and ancillary facilities was some £230 000 spent over 25 years. The original diffused air plant, extended in stages, cost £19 250 by 1960, while the mechanical aeration plant cost £46 200 in 1971. The 28 m diameter biological filters cost £37 770 in 1963. The sludge dewatering plant and electrical substation cost £61 000 in 1972, while miscellaneous equipment cost £51 000 over 25 years. Small sums were spent on the lagoon and pond, sludge-drying beds and stormwater tanks. The cost of removing 1 kg of BOD in 1974 was £0.15, including sludge removal from site and plant depreciation (as well as labour, maintenance, electricity, chemical and miscellaneous costs).

3.10.2 A petrochemical complex (ICI, 1976)

The wastewater from this plant is readily biodegradable, but experimental tests show that activated sludge treatment alone produces a sludge with a tendency to 'bulk'. However, the wastewater, with an average BOD approaching $1000\,\text{mg}\,l^{-1}$, is amenable to treatment by percolating filtration using a modular plastic medium. Furthermore, the effluent produced by preliminary treatment using high-rate percolating filtration can be further treated by conventional activated sludge treatment. The activated sludge can be settled satisfactorily and the final effluent is suitable for discharge to surface waters.

The high-rate percolating filtration stage consists of two stages, both modular plastic medium towers giving a total volume of $6600\,\text{m}^3$ of medium.

The plant is designed to accept $22\,800\,\text{m}^3$ per day of wastewater with an average BOD of $1000\,\text{mg}\,l^{-1}$. Reduction in BOD in the first stage is 60%, with a further reduction of 25% in the second stage. The overall reduction in BOD through the percolating filters is 70%, producing an effluent of strength approaching that of domestic wastewater for treatment by the activated sludge plant.

The emphasis in this example is on a simple, reliable method of primary BOD reduction, where fluctuations in BOD loading and possible shock loads of inhibiting substances can be more readily accepted without undue reduction in performance. It might well have been more cost-effective in the short term to employ only one high-rate percolating filter and to uprate the activated sludge plant to cope with a stronger influent. However, the second tower gives extra treatment capacity if required by future process plant expansion.

3.11 Conclusion

Wastewaters from the petrochemical industry contain a wide variety of substances which are potentially very damaging to the environment. Considerable emphasis has been placed on the removal of oils and immiscible solvents because they are visible forms of water pollution. However, serious long-term pollution and public health hazards can arise from toxic and persistent organic compounds in petrochemical wastewaters. The removal of these pollutants to very low levels may involve advanced wastewater treatment processes which are technically demanding and can be expensive.

Biological treatment methods are still favoured on cost grounds for treating strong organic petrochemical wastewaters and every effort must be made to avoid the introduction of toxic materials into such systems. More cost-effective ways of treating such effluents continue to be sought as pressure increases against the discharge of poor-quality effluents into estuaries and coastal waters.

With the ever-increasing costs of crude oil and base feedstocks for the production of petrochemicals, there is a strong incentive to conserve raw materials, products and byproducts at all stages of processing. Much can be achieved in reducing pollution at source by more rigorous attention to 'good housekeeping' in operating existing process plants. Full attention to pollution control problems at the design stage of new processes and process plants will continue to pay dividends in terms of cost-effectiveness in reducing pollution.

Acknowledgements

Thanks are due to Mrs Kathleen Gilbert and Miss Lynne Atkinson of the Centre for Extension Studies at Loughborough University for assistance with manuscript preparation and typing.

References

API, Division of Refining (1969) *Manual on Disposal of Refinery Wastes: Liquid Wastes*, American Petroleum Institute, Washington.
Bailey, D.A. (1976) The role of the regional water authority, *Chem. Ind.*, 808–818.
Bradfield, R.E.N. and Rees, C.P. (1978) The impact of toxic pollutants, *Effluent Water Treat. J.*, February, 61–71.
CEFIC Petrochemicals/Ecology Sector Group (1977) *Report on Task Force No. 2 on Treatment of Aqueous Effluents from the Petrochemical Industry*, CEFIC, Brussels.
Department of the Environment, Water Data Unit (1975) *River Pollution Survey, England and Wales, 1970 and 1975*, HMSO, London.
EPA (1980) *Treatability Manual*, Vol. III *Technologies for Control/Removal of Pollutants*, Report No. EPA 600/8-80-042c, US Environmental Protection Agency, Washington, DC.

Fielding, M. (1976) Oil pollution of inland waters, *Public Health Eng.*, 4(1), 18–22.
Flynn, B.P. (1975) A model for the powdered activated-carbon sludge treatment system, *Proc. 30th Purdue Ind. Wastes Conf.*, Ann Arbor Science, Ann Arbor, Michigan, 233–252.
Flynn, B.P., Robertaccio, F.L. and Barry, L.T. (1976) Truth or consequences: biological fouling and other considerations in the powdered activated carbon sludge system, *Proc. 31st Purdue Ind. Wastes Conf.*, Ann Arbor Science, Ann Arbor, Michigan, 355–362.
Hemming, M.L., Ousby, J.C., Plowright, D.R. and Walker, J. (1977) 'Deep Shaft' — latest position, *Water Pollut. Cont.*, 441–451.
ICI Pollution Control Systems Ltd (1976) *High-rate Biofiltration using Plastics Filter Media*, No. 5, *Chemicals and Some Other Miscellaneous Wastes*, Imperial Chemical Industries Ltd, Hyde, Cheshire.
Klein, L. (1962) *River Pollution*, II. *Causes and Effects*, Butterworths, London, 30–31.
Klein, L. (1966) *River Pollution*, III. *Control*, Butterworths, London, 101.
McCaul, J. and Crossland, J. (1974) *Water Pollution*, Harcourt Brace Jovanovich, New York.
National Water Council (1978) *River Water Quality, the Next Stage: Review of Discharge Consent Conditions*, London.
Nelson-Smith, A. (1979) The effect of oil spills on land and water, in *The Prevention of Oil Pollution*, Wardley-Smith, J. (Ed.), Graham and Trotman, London, 17–34.
Nounon, P. (1980) Fate and effects of oil in the marine environment, *Ambio*, 1X(6), 297–308.
Oil Industry Forum (1979) The estimation of oil in water with particular emphasis on production water discharges, *Report of The Oil Industry Internal Exploration and Production F rum*, London.
Oldham, G. (1979) Discharges from industrial plants into sewers, rivers and the sea, in *The Prevention of Oil Pollution*, Wardley-Smith, J. (Ed.), Graham and Trotman, London, 213–232.
Singleton, K.G. (1976) Methods and costs of industrial effluent treatment, *Chem. Ind.*, 233–237.
Stephens, H.K. (1970) Petroleum Chemicals, in *Our Industry Petroleum*, British Petroleum Co. Ltd, London, 321–333.
Wood, L.B. and Richardson, M.L. (1978) The water industry: safety and value for money, *Chemistry in Britain*, 14(10), 491–496.

4 Treatment of dyewastes

B D Waters, *Severn–Trent Water Authority*

4.1 Introduction

The process of converting raw, natural or synthetic fibres to finished materials makes use of large quantities of water and in many cases produces extremely polluting effluents. These effluents arise from the initial fibre preparation (e.g., scouring), bleaching, dyeing and printing, other finishing processes (e.g., mercerizing, application of flame-resistant finishes) and washing at various stages. Different fabrics receive different treatments and each stage of treatment produces a different type of effluent. Some effluents are readily biodegraded or can be treated by simple means; others may not be amenable to the conventional processes applied to most effluents that arise in other industries.

This chapter briefly reviews the background to the problems caused by wastes from dyeworks, with comments as appropriate, on the other effluents that may be present. The main part is concerned with describing the treatment alternatives available and how they may be used alone or in conjunction to treat the waste for discharge or for recycle.

4.2 Effluent standards

Two options may be available at a textile works with liquid waste for disposal. In many cases the effluent can be discharged to a sewer to be treated together with other domestic and industrial wastes at a municipal works. However, in many cases no sewer passes sufficiently close to a works so that effluent must be discharged to a river or a stream. This option may be available even though a sewer is nearby, giving the works operator a choice to find the more economic alternative. Sometimes, where the dual option exists, the local regulating authority may insist on disposal to sewer. In either case, restrictions are likely to be applied limiting the concentrations or load of contaminants that may be discharged.

In the case of discharge to river, limits will be applied to the biochemical oxygen demand (BOD), suspended solids and some other constituents such as metals and ammonia in order to protect the quality of the river. A similar range of restrictions may be applied to a discharge to a sewer although they will be less restrictive. Table 4.1 shows the type of limits that may be applied.

Table 4.1 Typical effluent standards for discharge to a watercourse or sewer (concentrations in mg l^{-1})

Actual consent conditions for particular discharges vary depending on the nature and flow of the effluent and the receiving stream. In most cases there are restrictions on the rate of discharge and the total daily flow to sewer.

Parameter	Concentration for discharge to	
	River	Sewer
BOD$_5$	20	
Suspended solids	30	500
Toxic metals:[a] Total	0.5	30
Soluble		10
Cyanide	0.1	5
Sulphate		300
pH value within range	pH 6–9	pH 6–10
Ammonia as N	10	

[a] Sum of concentration of, e.g., Cr, Zn, Cu, Cd, Sn, Pb.

In the USA an alternative approach has been proposed by the US Environmental Protection Agency (EPA, 1979) and Schaffer (1978) in which the polluting load is limited by regulations under the Clean Water Act. The proposed limits are based on restricting discharges to public sewers and to streams. It is proposed that the latter will be regulated by insisting on the adoption of 'Best Available Technology' by 1983. Limits are expressed in terms of kg per 1000 kg of fibre processed and are specific to each of a number of textile processes. They also vary with the size and type of textile mill. Table 4.2 gives some examples of the limits proposed.

Discharge to sewer usually involves paying the appropriate authority for treatment. Discharge to river is normally free but the discharger has to meet the cost of treatment himself. In some European countries there is a move towards charging for discharges to river on the basis of effluent quality, with incentives for improved quality (OECD, 1980). If there is a choice of disposal method the decision should be an economic one. However, in discharge to sewer the sewage treatment authority bears the ultimate responsibility for meeting the river discharge consent and the textile works manager is relieved of the problems of operating a full treatment plant. Even wastes discharged to sewer may need treatment in order to meet the consent of the sewerage agency. However, in this case the treatment plant will be less sophisticated and require fewer or less skilled staff to operate. Discharge to sewer also provides dilution by other wastes which may improve the prospects for treatment by reducing

Table 4.2 Examples of the regulations proposed in the USA for textile mill discharges (EPA, 1979)

Values given are the average of daily values for 30 consecutive days. Limits are also applied to the maximum for any one day. BCT is given for the smallest category of works — different values to larger works.

	Effluent limitations ($kg\,t^{-1}$ of product processed)		
	BAT[a]	NSPS[b]	BCT[c]
WOOL-FINISHING SUBCATEGORY			
BOD_5		8.9	11.2
COD	56.2	56.2	
TSS	6.4	6.4	17.6
Total phenol	0.018	0.018	
Total Cr	0.14	0.14	
Total Cu	0.14	0.14	
Total Zn	0.28	0.28	
Colour[d]	120	120	
pH range		pH 6–9	pH 6–9
CARPET-FINISHING SUBCATEGORY			
BOD_5		1.0	3.9
COD	16.3	11.2	
TSS	1.8	1.3	5.5
Total phenol	0.006	0.004	
Total Cr	0.02	0.02	
Total Cu	0.02	0.02	
Total Zn	0.05	0.05	
Colour[d]	220	120	
pH range		pH 6–9	pH 6–9

[a] Best available technology applicable to existing sources.
[b] New source performance standards.
[c] Best current technology applicable to existing sources.
[d] Colour measured in ADMI units (ADMI, 1973).

possible inhibitory effects on the biological processes in use at municipal works. Dilution also masks characteristics not readily removed, such as colour.

Attention is increasingly being paid to the possibilities for recycling effluents. Some effluents may be suitable for use as process-water without treatment; others may require treatment, depending on the quality of the effluent and the intended use. For example, wastewater derived from rinsing operations may be suitable, without further treatment, for scouring raw fibre. Wastewater already strongly coloured and of high pH value may be totally unsuitable for use in any process without extensive treatment.

In the overall operation of a works there may be opportunities to reuse effluents (for example, by reusing a dyebath after adjusting the strength) as well as to recycle for use elsewhere. Such a strategy can produce considerable savings in energy and chemicals as well as reducing the need for effluent treatment. However, some final wastewater always results and it is likely that extensive treatment will have to be applied. In many cases, the effluent from even a basic treatment plant (sedimentation, biological oxidation, secondary sedimentation and filtration) may be suitable for some purposes. The more exacting uses may require additional treatment such as adsorption by activated carbon or reverse osmosis. Where some treatment is already required before the effluent can be discharged to river or sewer, the additional cost of treatment to allow recycling is only that of the extra stages required. Some of this extra cost will be recovered in reducing water charges and effluent charges in many instances, making the concept even more attractive.

If water is in short supply there is an additional incentive to reuse and recycle. In these circumstances, costs are of secondary importance if production is at risk. Cost-effective treatment should still be the aim, however, and the choice of treatment requires careful consideration of the capital and operating costs of the alternatives available. Whatever the reasons for reuse or recycling of effluents, the opportunities are greater and the economics more favourable in a large, integrated textile mill. Although implementation will probably be more complex, it is likely to be worth the effort.

4.3 Problems posed by dyewastes

4.3.1 Impact on receiving water

The most obvious impact of the effluent from a dyeworks discharged to a watercourse is the colour that it often imparts. This can apply equally to a highly treated effluent as to a poorly treated one. A poorly treated effluent may also contain excessive amounts of oil, grease, detergents, other organic matter and suspended solids. All of these can have an adverse effect on the quality of a river. However, even if these materials are adequately removed, the residual colour can be more noticeable; this depends on the hue and concentration, bright reds being more noticeable than pale blues, for example.

Porter and Snider (1974) investigated the 30-day BOD of a number of textile-finishing chemicals including dyes and showed that there was little degradation or colour removal in many cases. They also showed (Porter and Snider, 1976) that the BOD represented only between 1% and 26% of the chemical oxygen demand (COD), depending on the chemical. This is less than the proportion for readily degradable compounds or domestic sewage (typically 50%).

The high ratio of COD to BOD is further increased by treatment if a biological process is used. The significance of this may depend on the other uses of the river. A high COD with a low BOD does not cause river deoxygenation but the presence of persistent organic compounds may adversely affect other water abstractors downstream. This is particularly true if water is being abstracted for the public drinking supply or for livestock watering. Fisheries can also be affected. Little *et al.* (1974) report the toxicity of 46 dyes to fathead minnows: overall, cationic dyes were found to be the most toxic, direct, vat and disperse dyes being the least toxic. The most toxic dye investigated was basic violet 1 with an LC_{50} of $0.047\,\text{mg}\,l^{-1}$.

ADMI (1973) also showed that 17 of 46 dyes investigated showed some inhibitory effect on at least one of four oxidative systems studied. Only two dyes inhibited all four oxidative systems. Two anthraquinone dyes also inhibited the anaerobic digestion of sludge. More recently, Brown *et al.* (1981) have described the development and results of a method for testing the inhibitory effects of dyestuffs on an aerobic wastewater treatment system. Using a standardized method developed for the Ecological and Toxicological Association of the Dyestuffs Manufacturing Industry, based on following sludge respiration rates, they showed that about 10% of the 202 dyestuffs tested may show an inhibitory effect if significant quantities reach a sewage treatment plant.

Fung and Miller (1973) have tested the inhibitory effects of 42 dyes on 30 bacterial species growing in solid culture media. They concluded that Gram-negative species are more resistant to inhibition than Gram-positive species and that basic dyes are more inhibitory than acid and neutral dyes.

Gardiner and Borne (1978) indicate the relative toxicities to fish of a number of textile-finishing chemicals. They found that the most toxic were dieldrin and DDT, both of which are declining in use. The least toxic were surfactants. The four dyes quoted exhibited toxicity of orders of magnitude similar to those of the metals copper, zinc, cadmium, nickel and chromium.

A matter of more recent concern with regard to persistent trace organic compounds that find their way into drinking-water supplies is their chronic toxicity — in particular, that they may be carcinogenic, teratogenic or mutagenic. Friedman *et al.* (1980) report that nine out of 28 dyes studied were mutagenic to the bacterial systems that they used. However, since they did not isolate the dye from the commercial preparation, they could not show whether it was the dyes themselves or the other substances present that were the mutagens.

Contrary evidence of mutagenicity was found by Rawlings and Samfield (1979). In a preliminary report of a study of 23 textile plant effluents, none was found to be mutagenic or toxic to rats at the maximum dosage given. The effluents could, however, be ranked in order of their toxicity to fresh water organisms. Evidence of the chronic toxicity of textile wastes is thus inconclusive at the moment as few and conflicting studies have been reported.

A number of dyes that were in use some years ago were subsequently found to be human carcinogens. In other cases, intermediates used in dye synthesis and/or trace contaminants of dye preparations showed similar properties. It is likely that in the present climate of concern many more studies will be carried out on the chemicals in use and on the effluents in order to clarify the risks to the workers handling them and to the environment.

It is established that dye wastes and other textile-finishing wastes can show toxicity to aquatic organisms although many such wastes are relatively innocuous. Their persistence is hardly surprising, as much research has been carried out by the manufacturers to develop stable dyes and other chemicals. The problems are compounded by the concentration of the industry in small areas. For example, in England cotton spinning and weaving are concentrated in Lancashire and the North West, the woollen industries are mainly in Yorkshire and the knitwear and hosiery industry in the East Midlands. Within these areas individual towns can have a large concentration of works and there are pockets of other textile industries scattered over the rest of the country (e.g., carpet manufacture at Kidderminster). In the USA 80% of textile mills are in the mid-Atlantic and Southern regions with the remaining 20% in three other regions (EPA, 1979). Similar concentrations of the industries can be found in most other countries. The consequences are that large mills or a number of mills discharge their effluent directly into the same river. Where textile works are discharging to sewer the textile wastes can contribute 50% or more of the total flow.

Textile mills also contribute other forms of air, water and solid waste pollution. These are outside the scope of this review. A general summary of the problems has been compiled by Ayers (1979).

4.3.2 Colour measurement

The recording of the true colour of water samples has always been a problem. Analytical methods are based on the absorption of light so interfering solids have to be filtered out before measurements can be made. Filtration of samples is often difficult as membrane filters of pore size <1 μm are used. Filtration also removes insoluble coloured substances that contribute to the perception of the colour of a waste.

Traditionally, measurements of water colour have been calibrated against a yellow–brown chloroplatinate standard at 465 μm (APHA, 1976). This is satisfactory for the natural colour of water due to dissolved organic acids, but unrelated to the spectrum of colours associated with dyeing. Consequently the American Dye Manufacturers' Institute has developed its own scale (ADMI, 1973). This measures the absorption over the range 400–700 μm and a calculation is performed which can be facilitated by the use of a computer.

The ADMI method has the advantage that it is independent of the hue but it requires more sophisticated instrumentation and more operator time to perform. Anthony (1977) has shown that for many effluents there is a reasonable correlation between the ADMI method and the APHA method. If a waste is not subject to a wide variation in colour and if such a relationship can be established, there is much to commend the APHA method for routine use.

4.4 The textile industry

The textile industry handles a wide range of fibres and manufactures a wide

DYEWASTES

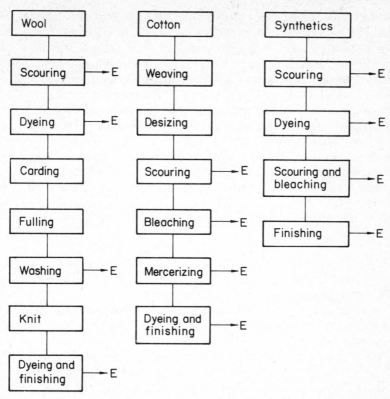

Fig. 4.1 Simplified processing sequence for some textile fibres (after EPA, 1974) E represents an effluent requiring treatment. Rinses are not shown

range of finished fabrics and other products. The main natural fibres processed are wool and cotton; hemp, jute and flax are examples of others less widely used. Increasingly the market has been dominated by synthetic fibres such as rayon, nylon and polyacrylics. Textile mills may be concerned only with the preparation or manufacture of fibres or may extend their activities into spinning, weaving, dyeing and finishing. Finished products include dyed and printed fabrics, clothing made from fabrics, carpets, knitted textiles and hosiery.

Figure 4.1 outlines the main steps involved in some of these processes and indicates where wastewaters may arise. In addition to the processes themselves, rinsing of the fibres or fabrics between process stages may give further volumes of wastewater. Details of the processing of fibres can be found in Trotman (1975) and Nemerow (1978). EPA (1974, 1979, 1980a) and OECD (1981) also give brief details together with information on water usage and the nature and strength of the wastes associated with each stage.

Treatment of effluents from textile plants has been described by Little (1967, 1975), Koziorowski and Kucharski (1972), Parish (1977), ADMI (1978),

McKay (1979a), EPA (1980a, 1980b) and OECD (1981). These studies tend to concentrate on the effluents associated with the early stages of textile processing — dyewastes are only briefly mentioned. This is understandable, as the wastes from the early stages are often the most polluting. Little has been published on the treatment of wastes arising from the manufacture of dyes, although Koziorowski and Kucharski (1972) do cover this topic. Dye manufacture is more diverse than textile manufacture but there are some similarities in the problems associated with the effluents. Both are coloured with dye residues and have a high content of other organic material. Much of this material consists of residues from the raw materials and the intermediates for the manufacturing process. Many of these are more biodegradable than the dyes themselves and treatment of the effluents will make use of the same types of treatment plant that are used for dyewastes. The advantages of reuse and recovery of materials as well as the recycling of water are equally apparent.

4.4.1 Preliminary textile-finishing processes

The preliminary processes used in textile manufacture and the nature of the associated effluents are briefly described because, in many factories, at least some of these are associated with the dyeing and final finishing stages. In such cases, effluent treatment has to be integrated at some stage. However, there may be advantages in isolating and treating some of the strongest or most difficult effluents individually before mixing them with the rest for the final treatment and discharge.

Wool is scoured in strongly alkaline solutions to remove dirt and grease. The resultant liquors have high BOD ($10\,000-20\,000\,\text{mg}\,l^{-1}$). The grease can be recovered by centrifugation or by neutralization and separation. The resultant waste can be treated by biological processes, preceded in many cases by coagulation and sedimentation. After scouring the wool can be dyed or bleached either before or after weaving.

Cotton is usually spun, woven into cloth and sized before it reaches the dyeing stage. Prior to dyeing it needs to be desized and then scoured (kier boiling) in alkaline solution to remove wax, dirt and grease. Both of these processes produce an effluent with a high BOD ($10\,000-20\,000\,\text{mg}\,l^{-1}$) that requires neutralization prior to conventional biological treatment.

Bleaching and mercerizing (a finishing process that gives a silk-like finish) of cotton produce wastes that are less strong than those from desizing and scouring. These can be treated by conventional means after mixing with other wastes.

Synthetic fibres need less pretreatment prior to dyeing as usually only sizing agents, antistatic agents and lubricating oils require to be removed. These processes are usually carried out at the dyeworks.

4.4.2 Constitution of dyewastes

It has been estimated that there are about 3000 dyes in use world-wide (McKay, 1979b) and that world production of dyes is 800 000 t annually (Anliker, 1977). Most of these products end up on the finished fabrics, but Anliker estimated that of the 360 000 t used in the textile industry, 10–20% was lost in the residual liquors. For all dye-using industries, about 47 000 t was lost to the environment even after allowing for 50% removal by effluent treatment.

Textile finishing makes use of a range of chemicals other than the dyes themselves. The total number of these is less and the problems of treatment are generally less. The main groups of chemicals are given in Table 4.3. Dyes can

Table 4.3 Some chemicals used in dyeing and finishing

Acids — inorganic and organic (e.g., formic and acetic)
Alkalis
Bleaches (chlorine, hydrogen peroxide)
Fluorescent whitening agents
Soaps and detergents
Dye carriers and other additives (e.g. o-phenyl phenol, benzoic acid, phenyl methyl carbinol)
Oils
Starch or substitute (e.g., carboxymethylcellulose)
Resins
Fire-, rot- and waterproofing agents
Pesticides
Silicates
Sulphides
Various inorganic salts
Organic solvents

be classified in different ways, according to use or chemical structure. For simplicity they are divided here into eight groups: Table 4.4 shows the main textiles for which they are used and the other chemicals often associated with their use. In many cases, the solutions are acidic or basic and in some cases metal ions are present. Sulphides are also used, together with various organic and inorganic compounds. Dyeing is frequently performed at high temperature, so the wastes may be warm even if heat recovery is practised.

Strength of dyewastes

The composition of wastes varies widely even on the same site, as different processes are usually conducted at the same works. Even a works that performs only dyeing produces wastes of differing strengths. Rinse-waters are relatively clean, whereas the effluents from scouring have high BOD and suspended solids concentrations. The BOD of mixed wastes from dyeworks typically varies between 200 and 3000 mg l^{-1}, with a COD between 500 and 5000 mg l^{-1}, suspended solids between 50 and 500 mg l^{-1} and a pH of 4 to 12. Many of the chemicals listed in Tables 4.3 and 4.4 may be present at varying concentrations. Individually the dyes and the carriers and other chemicals are usually measured in mg or g per litre.

Table 4.4 Main types of dyes, their main use and associated chemicals

Type of dye	Main use	Associated process chemicals
Acid	Wool, nylon	Sulphuric acid Acetic acid Sodium sulphate Surfactants
Azoic	Cotton	Metal salts Formaldehyde Sodium hydroxide Sodium nitrite Acids
Basic	Acrylic	Acetic acid Softening agent
Direct	Cotton, synthetics	Sodium salts Fixing agent Metal salts (copper or chromates)
Disperse	Polyester	Carrier Sodium hydroxide Sodium hydrosulphite
Mordant	Wool	Chromium and other metal salts Acetic acid Sodium sulphate
Reactive	Cotton, wool	Sodium chloride Sodium hydroxide Ethylene diamine
Sulphur	Cotton, synthetics	Sodium sulphide and other salts Acetic acid
Vat	Cotton, synthetics	Sodium hydroxide Sodium hydrosulphite and other salts Surfactants

The amount of dilution by rinse-water and other effluents (e.g., domestic waste from the factory) can have a significant effect and separation of waste to avoid dilution may be advantageous for some forms of treatment but disadvantageous for others. This aspect will be referred to again.

4.5 Treatment of dyewastes

4.5.1 Preliminary considerations

The variable composition of dyewastes, referred to above, can make control of treatment difficult. In addition, the flow can vary considerably, particularly if only daytime and/or five-day working is practised. This needs to be taken into account in designing a complete treatment plant. The design is also influenced by the intended means of final disposal. Disposal to sewer requires less treatment than disposal to river but the most sophisticated plant will be required if the intention is to recycle water.

There is scope to minimize the volume of effluent to be treated by maximizing the use of water by reuse and recycling. This has already been discussed. A distinction is drawn between reuse — using a dyebath several times by making up the strength each time — and recycling — using effluent from one stage (e.g., rinsing) for a less demanding use (e.g., scouring) without treatment first. The term 'recycling' is also applied to the use of treated effluent in place of fresh water supplies. The possibility of recycling treated effluents will be discussed as the description of the processes available is developed. At this stage, the main point is that reuse and recycling of untreated waste can not only save chemicals, water and energy but also reduce the size of the treatment plant needed. Of course, the waste that eventually results will be stronger and may demand larger units than would be required to treat the same volume of weaker waste. However, overall the plant costs (both capital and operating) should be less.

In the rest of this chapter it is assumed that full advantage has already been taken of heat recovery, solvent or other chemical recovery and reuse or recycling of untreated water.

4.5.2 Conventional treatment processes

Screening
If rags or other large solids are present, they should be removed as early as possible by screens. Mechanical raking can be incorporated to reduce the need for regular inspection. Failure to remove rags and long fibres results in the blockage of plant, particularly pumps and valves. They also wrap around mixers and other moving parts.

Flow balancing
Any treatment is simplified if flow balancing is achieved. The advantages are threefold. Firstly, the hydraulic load on the treatment plant can be kept fairly constant. Secondly, the chemical and biological load can be kept as constant as possible and temperature variations smoothed. Both of these advantages minimize sudden shocks to the treatment plant which can lead to a deterioration of the treated effluent. Finally, full benefit can be derived from mixing of acidic and neutral wastes in order to reduce the requirements for chemicals for neutralization.

In some cases, a further advantage is that toxic substances that can inhibit one of the treatment processes (e.g., metals in biological treatment) are diluted sufficiently that they are no longer a problem. However, using relatively clean effluents to dilute in this way may not be the most cost-effective solution if a much larger volume has to be subjected unnecessarily to a full treatment.

Where flow balancing is installed, good mixing is essential to get the best advantage. Close attention to the siting and design of the inlet and outlet helps to avoid short-circuiting. Mechanical mixing with paddles is required; alternatively, mixing can be achieved with aeration or pumped recirculation.

Neutralization

Strongly alkaline or acidic wastes need to be at least partially neutralized if discharged to sewer (see Table 4.1). Neutralization is also required for on-site treatment. In either case it can be carried out in the flow-balancing tank or as a separate stage. If both acidic and alkaline wastes are being produced, it is uneconomic to add both alkali and acid when the wastes can partially neutralize each other; individual neutralization should not be considered unless there is a good reason (e.g., separate treatment to recover a valuable residue or to precipitate an interfering metal).

Sulphuric acid is normally used for alkaline wastes. For large flows of acidic waste, lime is the cheapest chemical for neutralization but, for smaller flows, caustic soda is easier to handle. Despite the apparent simplicity of the process, neutralization plants can be unreliable. A continuous pH-monitoring system is usually essential and can be used to control the chemical feed. However, poor mixing, short-circuiting and the inability of the dosing plant to cope with rapidly changing pH values or flows can negate the advantages of a good monitoring and control system. Some of these problems can be overcome by neutralizing after flow balancing, either in a separate tank or in-line between the balancing tank and the next stage. Addition of the sulphuric acid or caustic soda by positive displacement pumps is the preferred method. Addition by drip feed or control by valve is usually unreliable.

Where wide variation in flow or pH cannot be avoided, neutralization may need to be carried out in two or more stages. In many works, alkaline wastes are neutralized by the addition of carbon dioxide in flue gas. Even if excess carbon dioxide is added, the pH will remain within the range acceptable for discharge to sewer. Details of this technique can be found in Little (1975).

Sedimentation

Depending on the waste and the subsequent processes, it may be necessary to include a sedimentation stage. If settleable solids are present, either derived from the materials being processed or resulting from precipitation after mixing or neutralizing the flows, these will be a source of blockage. This is a problem if the next stage is biological treatment or passage through a medium that can act as a filter (e.g., activated carbon). Sedimentation is usually carried out in a tank of proprietary design or in flat-bottomed tanks. These can be operated continuously or on a fill-and-draw principle. Proprietary designs usually give the most efficient and economic separation, but simple tanks are cheaper to install. Proprietary designs also include more satisfactory means to remove the settled solids so that frequent tank drainage is not required.

Residence time or flow rate through the tank varies with the nature of the solids and the degree of settlement required. Typical residence time is 2–6 hours. Upflow rates in proprietary designs of tank vary between 1.5 and 4 m h^{-1}.

Biological treatment

A biological stage is usually included as part of a complete treatment scheme. This is because of the biodegradable nature of many components of dyewastes and the fact that biological processes are generally cheaper than the alternatives to achieve the same degree of purification. Biological processes are a fundamental part of municipal sewage works treatment of dyewastes.

Both percolating filters and activated sludge plants are used and both can effectively reduce the BOD unless inhibitory substances are present. However, as previously discussed, many dyes and other textile-finishing chemicals are not readily degraded. There must be sufficient nitrogen and phosphorus present. In a sewage works treating a mixture of dyewaste and domestic waste this should not be a problem but in a plant treating only dyewaste nutrients may have to be added. Excessive concentrations of toxic metals, extremes of pH or the presence of other inhibitory substances can also have a detrimental effect. Standard texts on the use of biological processes for the treatment of wastes give guidance on the limits. A summary is given in Table 4.5.

Table 4.5 Conditions required to avoid inhibition of biological treatment

Temperature	≤35°C
Ratio BOD:nitrogen	~20
Ratio BOD:phosphorus	~100
pH value	range 6.5 to 9 (preferably ≤7)
Metals (Zn, Cu, Cr)	<10 mg l^{-1}

Percolating filters are widely used in sewage works and similar filters or high-rate filters (using plastic media) have been widely used for industrial waste treatment. Most reported treatment of dyewaste has been achieved with activated sludge plants. When activated sludge is compared with biological filters, the lower capital cost of the activated sludge plant and the opportunities for extended treatment and control outweigh the higher operating costs and sensitivity to shock loads. There can also be technical advantages (see Forster, 1977).

The low biodegradability of many dyes and textile chemicals means that biological treatment is not always successful, even with the extended treatment available in an activated sludge plant. The work of Porter and Snider (1974, 1976) who showed that, even after 30 days aeration, COD and colour were not much reduced has already been noted. Similarly, Michaels and White (1978) showed that conventional biological treatment plants reduce BOD but may not materially affect COD and colour. Dyes can be adsorbed onto activated sludge solids, however (Hitz et al., 1978; Dohányos et al., 1978), but this varies with the type of dye and operating conditions. Several workers have experimented with the addition of activated carbon powder to activated sludge plants in order to improve the adsorption of dyes and other materials. This is discussed in Section 4.5.4.

The non-biodegradable materials from wastes often pass straight through a sewage works so that the quality of the final effluent depends on the nature of the other wastes treated and the dilution. Wherever applied, biological treatment has its limitations but it is normally the most economic means of reducing the BOD as part of a complete scheme. There is a wide choice of alternative systems, including various derivations of the activated sludge process. Where site limitations are the dominating factor, high-rate processes such as high-rate filters or activated sludge plants will be favoured. Where land is available, simple lagoons with long retention times have been used. Lagoons have the advantage of requiring minimal attention and maintenance. They also use far less energy than activated sludge plants, but the high value of land often precludes their adoption in developed countries.

Biological processes depend on the growth of micro-organisms, which get carried over in the effluent stream. Separation in a secondary sedimentation tank is required. In the case of activated sludge plants, a proportion of the settled solids is returned to the aeration plant, but in all biological treatment plants there is some residual sludge that has to be removed. The sludges are generally difficult to dewater but may benefit from thickening and mechanical dewatering if they have to be removed off site. Sludge treatment is discussed further in Section 4.5.7.

Coagulation and sedimentation

Textile wastes may not contain much in the way of readily settleable solids. In order to improve the removal of settleable solids and to remove some of the colloidal components, coagulation can be employed. The usual coagulants are lime, aluminium salts (especially aluminium sulphate or alum), ferric chloride, ferrous sulphate or ferric sulphate. Synthetic polymers are increasingly used although, at present, they are more widely used as coagulant aids than as primary coagulants. New products are being developed continually and may find wider use.

Coagulation can be an effective means of reducing the colour of dyewastes and of reducing the BOD, COD and other components. Control of pH is critical in order to achieve the best results. This is especially true of alum, although satisfactory performance may be achieved over a range of pH values. Unfortunately, the coagulant and the dose that suit one waste may be totally unsuitable for another, so that the application of coagulation to a varying effluent can give mixed results. For this reason, flow balancing and mixing are important.

Dose rates can range up to several hundred $mg\,l^{-1}$ of alum or iron salt. Since these salts consume alkalinity, coagulant dosage needs to be considered for pH control. Alum coagulates best near to or just below neutrality. If the waste is strongly alkaline it is usually more economic and effective to use sulphuric acid to adjust the pH than to add excess alum. Iron salts can coagulate effectively at higher pH values and over a wider range so that pH control, with their use, may be easier.

The choice of coagulant, dose and pH can only be determined by a series of jar tests on representative samples of the waste. Where high concentrations of metals are present (e.g., copper and zinc), lime may be used to precipitate the metals and reduce colour (Netzer and Beszedits, 1975). Lime coagulation obviously takes place at high pH so that neutralization prior to coagulation is inappropriate.

All coagulant additions produce a significant quantity of waste sludge which is generally difficult to dewater and requires disposal. The use of coagulant aids (such as the synthetic organic polymers) and the optimum pH can help to reduce the amount of sludge by reducing the requirement for primary coagulant (Kace and Linford, 1975). Cationic polymers have been shown to be effective, on their own, for disperse dyes (Crowe *et al.*, 1977) and for water-soluble anionic dyes (Blank *et al.*, 1976). Similarly, a polyacrylamide has been used to coagulate a water-insoluble dye (Akhmedov and Garibov, 1966). Recent experimental work with magnesium salts as coagulants has shown some success (Fisons, 1976; Judkins and Hornsby, 1978) but their use has not been widely adopted on a commercial scale.

Coagulation–sedimentation is usually operated continuously using proprietary equipment. The tank designs used for sedimentation are usually of upward rather than horizontal flow and can cope with rise rates of between 2 and 5 m h^{-1}. Where the flow is low or discontinuous it may be more economical to operate static tanks on a batch basis. In this case the same tanks could be used to mix and neutralize the wastes prior to the addition of the coagulant. Higher-capacity mixers will be required than in a flow-through system. A similar approach may be advantageous where only a small part of the total waste flow requires this treatment, prior to mixing with the remainder for further treatment or discharge.

Flotation has been used as an alternative to sedimentation for the separation of coagulated solids, particularly where the floc has a low density. Higher flow rates are often possible so that the size of the separation stage can be reduced. The savings in capital costs on this part of the plant may be significant but there are the additional requirements of a flocculator, saturator and sludge collection equipment. As the plant contains a high proportion of mechanical equipment, energy and maintenance costs are higher than those for a simple sedimentation plant. A proportion of the flow has to be recycled through the saturator, increasing the energy costs further.

Flotation has only been adopted where the advantages are clear. Examples of this are grease or oil separation where the material being removed has a density less than that of water. Use of flotation for treating dyewastes seems to have been limited but at least one application has been described in which treated water was recycled and found suitable for dyeing (Boudreau, Dubeau, Lemieux, Inc., 1981).

It should not be forgotten that the chemicals used to coagulate wastes also remove phosphate. This may be considered an advantage if the waste is to be discharged to a watercourse liable to eutrophication. However, where biological treatment follows coagulation the phosphate may be required.

Filtration

The need for filtration is dependent on the quality required for final discharge or for additional treatment prior to recycle. In the former case, filtration would normally only be required to meet a particularly stringent discharge consent condition. For recycling, filtration may be required prior to further treatment by ion exchange, activated carbon adsorption or reverse osmosis. For these purposes, solids need to be removed in order to avoid fouling of the media or membranes in the subsequent stages.

Sedimentation tanks following chemical or biological treatment should produce an effluent containing less than $20-30\,\mathrm{mg\,l^{-1}}$ suspended solids, although this will depend on adequate design, operation and maintenance. If sedimentation has followed biological treatment, most of the suspended matter will be fine organic solids. Removal of these therefore reduces BOD and COD at the same time. Removal of suspended solids by filtration after chemical coagulation and sedimentation usually has little effect on BOD and COD concentrations. A well-designed and operated filter should reduce the suspended solids concentration in either case to below $5\,\mathrm{mg\,l^{-1}}$.

Filter performance depends on the design of the filter, the nature, quality and depth of the filter medium, the rate of filtration and the quality of the influent. The backwash regime and the method of control are governed to a large extent by the type of filter and medium employed, but they can also affect the way that a filter performs if not properly designed and operated. There are many different designs of upflow, downflow and crossflow filters. The first two of these can have more than one layer of medium. Most filters have to be shut down periodically in order to backwash the medium. There are also proprietary designs of filter that can be operated continuously, the medium being removed, washed and returned to the filter while it is in operation.

The choice of filter is dictated primarily by the quality of the influent, the quality required of the effluent, the desired throughput, the available head and the cost of alternative designs to meet these criteria. If river discharge is proposed and quality is less critical than for recycle, then the simplest filter may suffice. As the quality requirements become more rigorous, the degree of automatic control and monitoring of performance assume greater importance. The use of dual or multi-layer filters can improve quality or allow increased throughput.

Temperature, pH and biological activity on the filter medium can also have an effect on performance. Warmer temperatures favour the physical processes involved and biological activity is enhanced. The effect of pH depends on the nature of the solids and the medium; the normal range of operation is between pH 6 and 9. Biological activity on the medium helps to reduce both solids and soluble BOD.

The detailed theory of filtration is complex and beyond the requirements of this review. A useful introduction is provided by Jackson (1980a, 1980b) and details are given in most textbooks on water and wastewater treatment.

4.5.3 Polishing treatment processes prior to discharge

If river discharge is specified and there is a need to improve the quality of the final effluent, filtration is a suitable treatment to reduce suspended solids and BOD. It has little or no effect on dissolved colour and is comparatively expensive to install. Filters also require cleaning and regular maintenance. The techniques applied for the tertiary treatment of sewage may be suitable, especially if the main requirement is to remove suspended solids and BOD.

Microstraining
Microstrainers are finely woven drums of stainless steel fabric available in a range of mesh sizes. The drums revolve partially submerged in the effluent stream. As the liquid flows into the drum, solids are collected on the mesh. These are subsequently washed off by high-pressure jets and returned to an earlier stage in the treatment. Microstrainers may be cheaper to install and operate than filters but the removal of solids is not usually as good. Typical installations at sewage works remove 50% solids whereas a filter may remove 70%. The reduction of BOD obviously depends on the nature of the solids.

Land treatment
Where land is available, discharge to irrigation of the effluent from the final sedimentation tanks can be the cheapest method of improving the quality. The flow needs to be evenly spread over grass plots with a suitable fall in level. Loadings can be up to $8000 \, m^3 \, ha^{-1} \, d^{-1}$. As well as reducing solids and insoluble BOD, irrigation should give some reduction in soluble BOD and nitrification of ammonia. The degree of improvement in quality depends on the loading, the nature of the effluent and the climate. Colour removal is usually poor.

Grass plots need maintenance, particularly grass-cutting, and spare plots help to simplify this.

Lagooning
Storage of treated effluent in lagoons is also used for tertiary treatment. The lagoons are generally a metre or so deep and can be quite extensive in area, providing several days' storage. Suspended solids and BOD are removed by

sedimentation and biological action, and nitrification of ammonia takes place. Lagoons are more widely used in warmer countries such as Israel, South Africa and the southern USA. Although temperature plays an important role, the performance can be affected by short-circuiting within the lagoon or by turbulence due to excessive flows or wind action.

4.5.4 Adsorption processes

One of the aims of dye research is to increase the adsorption onto the fabrics being dyed. Consequently, adsorption of dyes onto suitable materials offers a means of removing them from effluents.

Activated carbon adsorption

The most commonly used adsorbent is activated carbon. This can be used in granular or powdered form and is made from a variety of materials including coal, coconut shell and wood products. The use of carbon derived from waste materials has also been investigated (Mitchell *et al.*, 1978).

The usual practice is to use granular activated carbon in columns. The depth of carbon is normally about 2 m, the diameter depending on the flow to be treated and the required contact time. Typical contact periods are from a few minutes up to 30 min or more. Maximum diameter is limited by cost considerations and practical design problems and rarely exceeds 2 m. Filter shells can also be operated in series as a means of increasing the contact. A further development of shells operated in series is the switching of the sequence so that the shell with the newest charge of carbon is always the last in the train, and the most exhausted carbon is first. This scheme optimizes the use of carbon.

An alternative arrangement is to use a moving bed in which fresh carbon is continuously added to the top of the adsorber in which the effluent is flowing upwards. Exhausted carbon is withdrawn from the bottom at the same rate that the fresh carbon is added. Figure 4.2 shows some of the configurations possible.

One of the attractions of the use of granular activated carbon is that the exhausted carbon can be regenerated for reuse. This requires a furnace and the operation needs to be carried out under carefully controlled conditions. The adsorbed material is carbonized and there is no waste for disposal. For small installations it is totally uneconomic to regenerate on site but the carbon can be sent back to the suppliers for regeneration. Provided that the conditions in the furnace are carefully controlled, the loss of carbon during regeneration should only be between 5 and 10%.

Granular carbon beds act as filters but the deposited material reduces flow and can inhibit adsorption. Some form of pretreatment is usually necessary in order to avoid this. Coagulation and sedimentation may suffice although a sand filter or other form of filter may be added as well. Because activated carbon, the

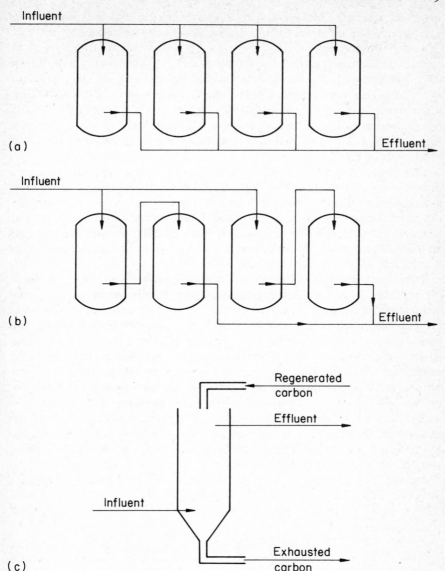

Fig. 4.2 Examples of configurations for activated carbon plants: (a) downflow adsorbers in parallel; (b) downflow adsorbers in series; (c) upflow adsorber with continuous carbon regeneration
In examples (a) and (b) part of the plant must be shut down in order to regenerate the exhausted carbon; plants of design (c) can be run continuously, the cleanest part of the flow being in contact with the most recently regenerated carbon

adsorbers and regeneration are relatively expensive, carbon adsorption is usually incorporated after the main organic load has been reduced by chemical or biological treatment.

Powdered activated carbon can also be used. The carbon is cheaper to buy in this form but cannot normally be recovered by regeneration. It is therefore wasted and requires disposal. Because the adsorption process is less efficient when powder is used, the amount used is higher. The powder can be added to flow-balancing tanks, neutralization tanks or separate tanks on a batch process. Alternatively, it can be added to a coagulation and sedimentation stage on a continuous or batch basis.

The choice of carbons is wide. A single manufacturer can offer a range of carbons derived from the same raw material which show different adsorption characteristics. Adsorption depends on pH, contact time, temperature and the nature of the substances being adsorbed. Different compounds are adsorbed best under different conditions (e.g., pH) and show different rates of adsorption on to different carbons. Choice of carbon and the design of the plant is complex if the waste to be treated varies in composition.

Preliminary selection of the carbons is usually made by laboratory measurement of the adsorption characteristics using samples of carbons and the compounds of interest. Such tests make use of pulverized samples of the carbon. Where comparisons are to be made it is important to control variables such as mesh size, temperature and pH. Carbons are characterized by their adsorption of phenol, iodine or methylene blue from solution under specified conditions. These measures can give an indication of the relative activity of carbons but may not always be appropriate for specific applications. They are not a substitute for laboratory studies with the compounds of interest.

The theory of the process of adsorption is complex. Details can be found, for example, in Weber (1972). A common model used in wastewater studies is the Freundlich equation which relates the residual impurity in solution at equilibrium to the amount adsorbed as follows:

$$\frac{x}{m} = kC^{n^{-1}}$$

or

$$\log\frac{x}{m} = \log k + \frac{1}{n}\log C$$

where x is the amount of impurity adsorbed
m is the weight of carbon
C is the residual concentration in solution
k and n are constants.

By varying the amount of carbon and the initial concentrations of the contaminants of interest, a range of conditions can be investigated with different carbons at various pH values and temperatures.

On the basis of these laboratory studies the most effective carbons can be selected for further investigation. In the case of mixed wastes, the best carbon for one component is rarely the best for all the components. Consequently, in order to complete the selection and to provide information on the optimum contact time and the life of the carbon (and hence to determine the size and flow rate of the full-scale plant), pilot trials should be conducted.

Under the right conditions most dyes and other organic compounds can be adsorbed. Disperse dyes are not well adsorbed, however, although high carrier concentrations can improve the removal at the expense of shorter carbon bed life (DiGiano and Natter, 1977). High-molecular-weight and non-polar compounds are more effectively removed than low-molecular-weight or polar compounds. The polarity of organic compounds is influenced by pH and this partially explains the pH-dependence of adsorption. Metals may also be adsorbed from solution and, as these are not removed from the carbon during regeneration, they can eventually reduce its activity. Other inorganic solutes are not adsorbed.

Carbon life to exhaustion is often longer than predicted by laboratory and pilot studies because of bacterial activity on the media. Bacteria oxidize adsorbed materials, allowing further adsorption to take place. This process has been referred to as biological regeneration. It can be encouraged if a small amount of municipal sewage is present (Schwägler and Stotz, 1980).

The combined process of adsorption and biological oxidation has been used to advantage by adding powdered activated carbon to activated sludge plants. The residence time of the carbon in the plant is several days (determined by the sludge wastage rate) and it eventually becomes saturated with non-biodegradable materials. It should be possible to balance the input of fresh carbon with that removed with the surplus activated sludge. Carbon concentrations in the plant are several hundred $mg\,l^{-1}$, making the cost high. This process has not apparently found much application in treating strictly dyewastes although some work has been reported.

Bettens (1979) describes an application in a Belgian plant treating wastes from the dyeing and printing of wool and nylon at a carpet dyeworks. Residues from wool washing, detergents, thickening agents, biocides and various other contaminants were present with the dyes. The existing plant consisted of flow balancing, activated sludge, dual media filtration and adsorption on activated carbon. Although the treatment plant produced an effluent of adequate quality, the running costs, especially those for the regeneration of the carbon, were high. There were also operational problems due to varying load or to adverse reactions to one of the constituents of the waste.

Powdered activated carbon ($100\,mg\,l^{-1}$) was added daily to the activated sludge plant until the concentration reached $1\,g\,l^{-1}$. Eventually, they found it necessary to add $50\,mg\,l^{-1}$ on a daily basis in order to maintain the performance. By comparing equivalent periods prior to and after the addition of the carbon, the COD of the effluent from the activated sludge plant was found to be reduced from $181-374\,mg\,l^{-1}$ to $117-140\,mg\,l^{-1}$ and colour was reduced from 350–990 to 189–296 APHA units. BOD was also reduced and nitrification improved. As a consequence, the reliability of the plant was improved and water could be discharged to waste after filtration. Granular activated carbon was used to treat only the water recycled through the works.

Voorn (1976), using parallel pilot plant studies, showed that the daily addition of $100\,mg\,l^{-1}$ of powdered activated carbon to an activated sludge plant improved COD removal from 80% to 84%, TOD from 52% to 59% and colour removal from 46.5% to 82%. This work was undertaken with an unspecified textile waste of initial COD of $1181\,mg\,l^{-1}$ and TOD of $1093\,mg\,l^{-1}$ (colour not specified).

Further details of the design of plants and the testing and applications of carbons in treatment plants are given by Cheremisinoff and Ellerbusch (1978) and De John (1976).

Adsorbents other than carbon

Activated carbon is the most widely used adsorbent in effluent treatment but a variety of other materials have been used. Basic dyes have been shown to be adsorbed on to silica gel (Alexander and McKay, 1977) and fuller's earth (Thornton and Moore, 1951). Dyes are also adsorbed by activated alumina (Mutch, 1946), peat (Leslie, 1975), wood (Poots *et al.*, 1978), clay (Sethuraman and Raymahashay, 1975) and synthetic polymers (Ciba Geigy, 1975). Metals and other organic compounds are also adsorbed.

The relative merits of these different materials have been compared by McKay *et al.* (1978, 1980a, 1980b) and Poots *et al.* (1976a, 1976b), who showed that activated carbon had the highest capacity for most of the dyes studied. Carbon was more expensive than peat or wood (but cheaper than silica) so it may not be the most cost-effective. The ability of carbon to be regenerated might reduce the long-term cost. Peat and wood would not be regenerated but could be burnt to produce steam.

These alternative adsorbents are unlikely to be widely used unless they are found both uniquely suitable and available locally at favourable cost.

4.5.5 Chemical oxidation

Chemical oxidants can destroy the colour of dyes and oxidize other materials. The oxidation may not be complete but the oxidation products may be more biodegradable and therefore amenable to conventional treatment. A number of oxidants have been used in experimental plants and on the full scale.

Chlorination

Chlorine and sodium hypochlorite are effective in decolorizing some dyes and have been used for the treatment of effluents from the production of dyes (Éndyus'kin *et al.*, 1979). Both chemicals are relatively cheap and are often used in textile plants and chemical works for other purposes. Oxidation with chlorine may not be environmentally acceptable, especially if the waste is discharged directly to a river. Excess chlorine can be controlled but the chlorinated products may be as undesirable or their effects as unknown as the dyes themselves.

Ozonation

More interest seems to have been shown in the use of ozone. Ozone is one of the strongest of the oxidizing agents that are likely to be used in an effluent treatment plant. (Only hydrogen peroxide in the presence of ultra-violet radiation could be considered a stronger alternative oxidant.) Ozone has to be generated on site from a dry supply of air or oxygen: the use of oxygen improves the yield of ozone but at higher cost. The gas is passed through a generator in which it is subjected to a high-voltage electric field. The generator is expensive and power consumption is of the order of $20-25\,W\,g^{-1}$ ozone produced. The ozone must be added to the wastewater with good mixing to ensure full use of the ozone before it bubbles to the surface of the contact chamber. This means that deep tanks are required, usually $2-4\,m$.

The solubility of ozone in water is proportional to its partial pressure, hence increasing the yield of ozone from the ozonizer increases the quantity of ozone that passes into solution. Increasing the total pressure by having deep tanks and working at the lowest practicable temperature also increase the overall efficiency. Pure ozone is over ten times more soluble in water than oxygen but at the ozone dilutions normally obtained from an ozonizer (typically $1-2\%$ in air) the solubility at $10°C$ and 1 atm pressure is about $10\,mg\,l^{-1}$. Adding higher doses to water at normal pressures requires the continuous addition of ozone to replace that used up in the oxidation reactions.

Other factors that affect the overall efficiency include bubble size and the degree of mixing. Contact time is also important as the gas is relatively unstable. The half life of ozone in pure water is about 25 min at neutral pH. Higher pH values catalyze the decomposition. Hence, the rate of reaction with oxidizable material depends on the interrelationship of several variables affecting not only the rate of substrate oxidation but also the rate of ozone decomposition.

Ozone is a toxic and corrosive gas and the residual cannot be safely vented from the reactors if there is plant or personnel nearby. The reactors have to be enclosed and provision made either to recycle the gas or to destroy the residual ozone. Destruction can be accomplished by passing the vented gas over granular activated carbon or a metal catalyst supported on alumina. In the latter case, the temperature may need to be raised slightly. Heating the waste gas to $300°C$, in the absence of a catalyst, will also destroy ozone.

Fairly high doses of ozone are required for the decolorization of dyewastes. Reported doses range from $45\,mg\,l^{-1}$ (Nebel and Stuber, 1976) to $1000\,mg\,l^{-1}$ (Snider and Porter, 1974). There is usually only a slight reduction in COD or BOD but colour removal can be complete. In view of the high cost, ozonation is usually applied as a final treatment after most of the potential demand has been reduced. Erndt and Kurbiel (1980) showed that the application of ozone in more than one stage was more effective than applying the same dose of 150 mg l^{-1} at once. One stage required 30–40 min contact to remove 55% of colour and 30% of BOD, whereas the two-stage treatment removed 67% of colour and 85% of COD after a total of 15 min contact. Reduction of the concentrations of anionic and nonionic detergents was similar in both cases at about 85% and 60% respectively.

More efficient use of ozone has been claimed for the simultaneous use of ultra-violet (UV) radiation in colour removal from kraft mill effluent (Hobson, 1977). Although the addition of the UV equipment increases the capital cost, this may be more than offset by the savings in ozone. This variation has not yet been widely reported for the treatment of dyewastes but may find wider application in the future, particularly if ozone itself becomes more widely used.

The use of other oxidation techniques has been reported. Rohrer (1977) reports the use of oxygen with a catalyst to decolorize a textile-finishing waste in two steps. An initial oxidation decomposed stabilizers and complexing agents, facilitating the second step of coagulation and sedimentation. A final oxidation step produced an effluent that was suitable for recycling. Sidwick and Barnard (1981) report use of oxidation, catalyzed by manganese chloride, as a pretreatment for sulphide removal prior to high-rate biological treatment of a waste derived from the manufacture of canvas cloth. Similar processes based on the 'Katox' process have been described by Hocke (1978).

4.5.6 Other treatment processes

Ion-exchange resins and macroreticular resins have been used to remove dyes from solution. Jørgensen (1974) used a cation cellulose exchanger followed by an anion-exchange resin after pretreatment by coagulation with alum and sedimentation. The exchangers were regenerated with caustic soda and the treated water was reused for dyeing. Removal of polyamide, polyester and mixed dyes was better than 93%.

Rock and Stevens (1975) investigated treatment with a macroreticular resin followed by a weak anion resin. They regenerated with methanol although the anion resin had to be pretreated with caustic soda and reactivated with sulphuric acid. The advantage of methanol regeneration was that it could be distilled off and reused, leaving a concentrated residue for disposal. Acidic, basic, reactive and direct dyes were effectively removed but the removal of disperse dyes was not as effective. Colour removal was 93% from an influent with a colour of 1160 APHA colour units (APHA, 1976) to over 98% from an influent of 475 APHA colour units. An economic analysis for one dyehouse showed that the capital cost of plant to reduce the colour of a 189 m^3 d^{-1} effluent from 1000 APHA to 50 APHA units would be 40% of that of an equivalent activated carbon plant and the running costs would be 70% less.

DYEWASTES   215

A similar process was investigated by Maggiolo and Sayles (1977). About 90% colour removal was found with an acid dye, a reactive dye and disperse dyes. Removal of COD was generally less than this and it deteriorated with throughput faster than colour removal. Methanol was used for regeneration and, overall, the costs were estimated to exceed or equal those for ozonation, activated carbon or reverse osmosis at a similar scale (76 m^3 d^{-1}). They claimed that the process would be more effective than these alternatives for the removal of disperse dyes.

Reverse osmosis has been suggested as a treatment process. Working with cellulose-acetate-based membranes on pilot scale, Cohen (1975) showed that pretreatment by coagulation with ferric chloride reduced the rate of decline of the flux rate. However, even without pretreatment, removal of colour and COD was good. Brandon (1980) reports full-scale trials with zirconium oxide–poly-acrylic acid membranes on stainless steel tubes in which the recovered water was reused. Up to 95% water recovery was achieved after using a variety of dyes. The water was satisfactorily reused as process-water. The effluent was treated at 85°C, allowing savings in heating such that the estimated payback period would be 5.2 years. Initial investigation showed that the concentrate might be reusable and it was thought that, if this was developed, the payback period would be reduced to 3.8 years.

A treatment method involving solvent extraction has been patented (Kitamura *et al.*, 1979) but seems to be of rather specialized application.

4.5.7 Sludge disposal

Sludges and solid wastes may arise from a number of stages in a treatment plant. These are summarized in Table 4.6. It is often the case that an effluent

Table 4.6 Solid wastes arising from dyewaste treatment

Treatment process	Nature of solids
Screening	Rags, paper, fibres
Neutralization	Precipitated solids, metal hydroxides
Coagulation and sedimentation	Aluminium, calcium or iron sludges
Biological oxidation	Surplus activated sludge or humus solids
Filtration or adsorption with activated carbon	Filter wash-water

treatment plant produces a good-quality effluent but presents problems to the operator only because little thought was given to sludge disposal.

Chemical coagulation and biological oxidation give rise to large volumes of sludge that usually contains less than 1% solids. If the plant is operated in order to reduce the pollution load to river or sewer, this sludge requires separate disposal. If the plant is used to recover water for recycling, it may be possible to dispose of the sludge to a sewer. However, if the solids concentration is too high the sludge may not be accepted and, in any case, the charges may make this uneconomic.

Materials removed by screens are mainly large solids such as rags, fibres and paper. These are readily dewatered to a handleable state and can be tipped or incinerated. Sludges produced by lime precipitation are usually more readily dewatered than those from aluminium or iron salts but all chemical and biological sludges normally require thickening prior to further treatment. Wash-water from filters and activated carbon plants can be similarly thickened, although it is common practice to recycle wash-water back to an earlier sedimentation stage in the treatment plant. Similarly, secondary sludges from biological treatment plants may be recycled to an earlier sedimentation stage for the first stage of concentration. At some stage, all the sludges will need to be transferred to a thickening tank prior to further dewatering.

Mechanical dewatering with filter presses, centrifuges, belt presses, vacuum filters or similar plant may be used. Most sludges require chemical conditioning with polymers, lime or other chemicals prior to feeding to the dewatering plant. If the conditioning is right and the choice of equipment is suitable, it should be possible to dewater sludges to 20% solids or better. The choice of plant and conditioner depends very much on the nature and volume of the sludge and may best be finalized after the effluent treatment plant is operating and samples of sludge are available for laboratory and pilot plant trials.

The volume of sludge can be estimated from the chemicals used and the pollutional load removed. Laboratory and pilot plant results for the effluent treatment may be useful in this respect. However the design is arrived at, sludge treatment needs to be considered at an early stage in the design process so that at least some provision is made to store and thicken the sludge. Land has to be made available if the plant is to be installed after the treatment plant and the costs of sludge disposal have to be included in the economic and financial appraisal.

4.6 Complete treatment plant design

4.6.1 General principles

Design of a complete treatment plant depends on a number of factors. The starting-point is often the final discharge, i.e., where can the final waste be discharged and what are the conditions to be observed? If a sewer is available, the in-house treatment plant will probably be less complex than that required for a discharge to a river. Discharge to sewer usually incurs a charge related to the volume and strength of the waste so that the degree of treatment applied depends equally on economic factors as on meeting the terms of the consent. Where there is a choice, the relative economics have to be compared for the different degrees of treatment required, taking full account of the running costs. The larger and the more complex the treatment plant, the greater the problems of control and the greater the requirement for skilled labour. This escalation can make a significant contribution to the total running costs.

A reliable treatment plant cannot be designed purely as an exercise on paper, following the experience of others. Even for the most basic plant design, representative samples of the effluents and the mixed effluents are required. Analysis of these can be used to determine the likely range of effluent qualities that have to be designed for and to estimate the size of balancing tanks and pumps required. They can also indicate the requirements for plant and chemicals for neutralization.

Designing more sophisticated treatment normally requires laboratory tests to determine the best coagulant or other chemical treatment to be used and the costs likely to be met. A comparison may then be made with other forms of treatment, for example, biological treatment for BOD removal, based on outline designs of potential treatment schemes. Further work with pilot plant is often used to confirm the results of the laboratory studies and to provide further design information. Pilot plant trials are essential if activated carbon plant is to be installed or if more complex treatment is under consideration (e.g., powdered activated carbon in activated sludge).

It is important to consider the flexibility of the plant to cater for a range of flows and qualities of effluent at an early stage. Variation may be due to the daily working pattern, to seasonal changes in production or to developments in dyeing technology. It should be remembered that biological treatment plants have to be run continuously, they cannot be shut down for several days and then started up again and expected to perform well. Similarly, coagulation and sedimentation plants that are designed for continuous operation do not respond well to sudden variations in flow. Factors such as these will affect the amount of flow balancing required and may rule out some particular pieces of equipment.

In many works, all of the effluents are drained to a common point for treatment. This may be disadvantageous if chemicals are to be added and there are components of the flow that are relatively clean. Diluted effluent may require less chemical, but the reduction is not usually in proportion to the dilution. On the other hand, dilution may be beneficial if biological treatment is proposed, from the points of view both of reducing high organic loadings on the plant and of diluting inhibitory substances. It may be more economical to keep some of the waste streams separate from the stronger wastes and treat them for recycling, rather than trying to treat the whole flow to a suitable quality.

Separate treatment may also be appropriate for wastes containing sulphides or metal salts. Sulphides occur mainly in wastes from the use of sulphur dyes (Table 4.4). Aeration at acidic pH values can reduce the concentration of sulphide in solution. However, the resultant hydrogen sulphide gas needs treatment in its own right, so this is rarely an attractive proposition. The usual alternatives are oxidation with hydrogen peroxide or precipitation with iron salts and lime. Oxidation can also be carried out with chlorine or by catalytic processes but, in any case, is expensive if there is a strong oxidant demand from other constituents of the waste. Precipitation may demand large quantities of chemicals and produce excessive volumes of sludge. Dilution and treatment with the rest of the waste is the most economic approach if the sulphide does not interfere with treatment and if the final concentration is acceptable.

Wastes containing metal salts are also worth considering for separate treatment as they are not adequately removed by conventional processes and can interfere with biological treatment. The precipitation of copper by increasing the pH to 8–9 with lime and the precipitation of chromium by ferrous sulphate and lime result in a waste that can be settled. The supernatant liquid phase may be mixed with the rest of the flow. Ultimate settlement can even take place in a subsequent part of the waste treatment plant, if contact with the rest of the flow does not cause dissolution of the precipitated metal hydroxide.

4.6.2 Discharge to sewer

The minimum treatment required for discharge to sewer may be flow balancing and screening. If the pH value is outside the consent limits, neutralization is required. If other limits for toxic metals, sulphides, organic solvents, suspended solids, COD or BOD are exceeded, further stages of treatment are required. This may take the form of chemical addition (e.g., coagulation) and sedimentation, or biological treatment and sedimentation. Rarely are chemical and biological treatment applied, nor are additional forms of treatment required.

4.6.3 Discharge to a watercourse

Increasingly, more stringent conditions are applied to discharges to rivers and first-time controls are being applied in areas where none have been applied before. Consequently, discharges to rivers are being treated more extensively than in the past and to a higher degree. Biological treatment is often the only treatment used, although this requires final sedimentation and may also require preliminary sedimentation.

Coagulation and sedimentation may be applied to the waste before biological treatment in order to reduce the load, to remove or reduce the concentrations of toxic or non-biodegradable matter or to be an alternative to the required sedimentation without coagulation. In some cases biological treatment precedes coagulation. This may be useful if much of the organic load in the waste is readily degraded; relatively low chemical doses are then sufficient to remove non-degraded substances.

Separation of the wastes to limit coagulation to those wastes that require it has already been mentioned. This may even be carried out in a batch process. If coagulation is carried out with lime at high pH, adjustment to the range 7 to 8 will be required prior to biological treatment.

Other forms of tertiary treatment may be required after biological treatment.

Processes commonly used include lagooning or running over grass plots; these require large land areas and are not usually practicable on industrial sites. They may find use at municipal works receiving dyewaste. Straight filtration through a sand filter or microstraining may be used to reduce suspended solids and BOD even further. Where colour removal is required and cannot be achieved by coagulation or biological processes, ozonation or activated carbon treatment may be required.

4.6.4 Recovery of water

Effluent can be treated to a quality suitable for reuse in the full range of operations carried out at most textile works. The more sensitive the water quality demands of the operation, the more treatment is likely to be required and hence the greater the cost. If the water is to be reused for dyeing or final rinsing, some form of tertiary treatment will be required in most cases. For instance, sand filtered and biologically treated sewage effluent (containing the textile mill waste) has been found to be acceptable for recycling (Harker, 1980).

An alternative to coagulation, biological treatment and chemical oxidation or adsorption is to adopt one of the less common processes such as reverse osmosis or ion exchange. In most cases, some form of pretreatment will be required in order to avoid blockage or fouling. Filtration may be all that is required but, as reverse osmosis and ion exchange are relatively expensive to install and operate, there may be some advantage in reducing the load to these stages by prior chemical or biological treatment.

Reverse osmosis separates water from the dissolved impurities by passage under pressure through a membrane. This means that, in addition to the purified water, a more concentrated effluent is produced for disposal. The volume and strength of this residual effluent depend on the strength of the influent, the membrane used and the operating characteristics of the plant. Volumes are typically 25% of the influent flow. This effluent may require treatment before it can be discharged. Return of this effluent to an earlier stage of treatment may be practicable but, as reverse osmosis is probably the only treatment removing the contaminants concentrated in this waste stream, separate treatment is more likely.

Ion-exchange resins (and resins that adsorb organic substances by non-exchange processes) are not regenerated by heating to high temperatures like activated carbon. Instead, they have to be treated with a suitable solvent which removes the contaminants from the resin, leaving it ready to use again. The regenerant may be aqueous (e.g., caustic soda) or an organic solvent (e.g., methanol). The choice depends on the type of resin and the nature of the contaminants. In either case, the result is a concentrated solution of the original waste. Regenerant volumes are typically a few percent of the treated effluent flow but the volume depends on how easily the contaminants are removed by the regenerant. Ion exchange is generally used to treat weak wastes —

otherwise the volume of regenerant approaches the volume of waste treated. If the regenerant is an organic solvent, it can be recovered by distillation; alternatively, it may be economic to incinerate the regenerant solution. Strongly acidic or alkaline regenerants require neutralization during which much of the soluble material should precipitate. The residual solution may still require some treatment, perhaps by return through the plant.

4.7 Summary of treatment processes

Any of the processes described above can be fitted into a treatment scheme, although it is unlikely that more than two or three together would be required for discharge to sewer. Six or more stages of treatment might be required in order to obtain water of a quality suitable for recycling. Figure 4.3 outlines the options available to meet different requirements in the order in which they are likely to be applied. This order may well have to be changed to suit particular circumstances.

Some stages of treatment produce wastes of their own, e.g., sludges or regenerant solutions. Their disposal has been discussed already. Where recycling is to be practised, separation of effluents and the fullest use of untreated effluents may significantly reduce the size of the treatment facilities required.

4.8 Case studies

There are many published reports of studies in which various treatment options have been investigated and compared for removal efficiency. Many of these are laboratory or pilot plant studies directed toward selection of the best process or combination of processes. Rarely can processes be compared on a full scale. In the majority of reports little or no cost information is given.

It is a general rule in effluent treatment studies that no two wastes are the same. This is particularly true of textile-finishing wastes in view of the variety of fabrics and possible finishing processes. Despite the limited number of treatment unit processes available, a surprisingly large number of different combinations can be found.

Some detailed case studies involving discharge to sewer or stream or reuse have been presented by Atkins and Lowe (1979). These authors investigated a number of works dyeing and finishing wool, cotton and synthetic fibres. After discussing these they draw attention to the increase in running costs as well as capital costs as the plants become more sophisticated. They also point out the economies of scale in treatment plant which make pollution control relatively expensive for the small works.

In the succeeding sections other examples are quoted from the literature.

DYEWASTES 221

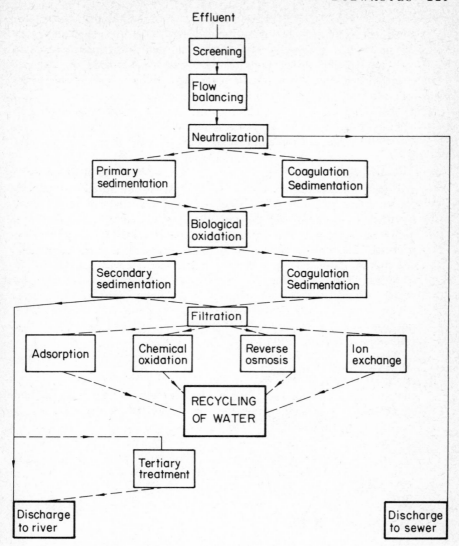

Fig. 4.3 Options for treatment of effluents for discharge or recycling

The examples are not exhaustive, but have been chosen to give a cross-section of the schemes studied or installed. Costs are given in the currencies and prices quoted in the original paper.

(1) Rinker and Sargent (1974) describe in some detail a plant treating waste from the dyeing of various fibres by flow balancing, activated sludge, alum coagulation and chlorination. Screening had already been carried out

prior to heat recovery. Some settlement occurred in the flow balancing tanks. Ammonia was added to the activated sludge plant but phosphate was not required. The effluent from the activated sludge plant was settled and some of the sludge recirculated. Alum, caustic soda, a polymeric coagulant aid and hydrogen peroxide were added in a mixing basin, followed by settling in a clarifier. The final effluent was chlorinated prior to discharge to a stream (chlorination of effluents for disinfection is common practice in the USA but is not widely used elsewhere). Excess activated sludge and the alum sludge were dewatered by centrifuge, the centrate being recycled to the flocculation basin.

Overall reduction of BOD was 93%, of COD 73% and of colour 69%. Further work to improve the system included a two-step alum coagulation. This was carried out initially at pH 5. The pH was adjusted to 6.5 prior to filtration. Removal of acid and disperse dyes was optimized at the lower pH value; metal and phosphate removal occurred at the higher pH. These measures added $200 000 to the capital cost. Colour removal was not considered adequate, however, and ozone, activated carbon and resins were investigated in pilot plant as alternative additional treatment processes. Ozonation was selected and the final capital cost to treat $3700 \, m^3 \, d^{-1}$ of wastewater (arising from the processing of $37 \, t \, d^{-1}$ of fibre) was estimated to be $1 600 000. Annual operating costs were estimated to be $369 200 (prices 1974).

(2) Mahloch et al. (1974) started their investigation by looking at the biological treatment of a scouring and dyeing waste, including the addition of $200 \, mg \, l^{-1}$ activated carbon to a pilot activated sludge plant. These were subsequently compared with lime coagulation and activated carbon adsorption.

The activated carbon improved the performance of the activated sludge plant only marginally. In both cases BOD reduction was 90–95% and COD reduction was 50–55%; colour removal was poor. Adsorption with $3000 \, mg \, l^{-1}$ powdered activated carbon followed by coagulation with $670 \, mg \, l^{-1}$ was judged to be too expensive.

(3) Singer (1976) investigated 20 dyewastes for the American Dye Manufacturers Institute. He concluded that, whereas BOD was effectively reduced by biological treatment, colour was not. Disperse, vat and sulphur dyes were decolorized by coagulation with alum but not by adsorption with activated carbon. Reactive dyes were most effectively decolorized with ozone but, in common with basic, acidic and azoic dyes, they were also decolorized with activated carbon. Disperse dyes were the least effectively decolorized by ozone. BOD and TOC were poorly reduced by all of the physicochemical techniques.

(4) Shelley et al. (1976) ran parallel studies with an unspecified dyeing and textile-finishing effluent, comparing coagulation with either 4500 mg l^{-1} lime or 600 mg l^{-1} alum followed by neutralization and biological treatment and filtration. The alum stream achieved 54–69% removal of COD and 72–87% removal of colour, whereas the lime stream gave reductions of 53–80% and 77–87%, respectively. Most of the colour was removed by coagulation. Comparison of these results with a similar parallel trial in which coagulation with alum or lime was followed by adsorption with activated carbon gave similar overall performance. Lime again appeared to give the better result; however, in both of the trials using lime, the high doses required, the consequent chemical demand for neutralization and the larger volume of sludge produced made lime treatment comparatively the more expensive method.

(5) A French report (Anon., 1977) and a German paper (Freytag and Diemunsch, 1979) emphasize the need to reduce waste volume as low as possible and maintain concentration as high as possible — not only to reduce costs but also to make treatment practicable. These studies compared treatment of cotton-finishing wastes by a number of processes including coagulation and sedimentation or flotation, adsorption on activated carbon, catalytic oxidation, activated sludge and ozonation. The objective was to reduce pollution by 95% and colour to a maximum residual of 100 mg l^{-1} APHA units. Coagulation and sedimentation had the disadvantage of producing voluminous quantities of sludge. Activated carbon adsorption required pretreatment and reduced COD by only 50% despite the low throughput of two bed volumes per hour. Catalytic oxidation also required pretreatment and was found to be not very effective or reliable. The recommended treatment for effluent not containing chromium or sulphide was activated sludge including a small dosage of alum. Expected reduction of COD was 80% and of BOD 90%. Residual colour removal would require an additional stage such as ozonation.

(6) Shirazi et al. (1977) in Iran investigated first-time treatment for a dyeworks using a mixture of dyes. They found that coagulation with ferric chloride reduced COD by 51% and colour by 68%. Inclusion of activated sludge treatment increased the overall reduction of both COD and colour to 75%. Activated sludge on its own reduced COD and colour by 63% and 51% respectively.

(7) Murakami and Annaka (1978), in Japan, studied the removal of colour from municipal sewage containing about 10% mixed textile wastewaters by activated sludge followed by tertiary processes. Activated sludge removed only 30–40% of dissolved colour but insoluble colour was

almost completely removed. Addition of $10\,\text{mg}\,\text{l}^{-1}$ ozone removed 60–70% of the soluble colour. Adsorption by activated carbon varied depending on the dyes present and the carbon used. Coagulation was not considered suitable as a tertiary process.

(8) In the UK, Waters (1979) reports the use of coagulation of a dyewaste prior to mixing with the main flow to a sewage works. The dyewaste represented about one-third of the total flow and was previously treated by lime addition ($500\,\text{mg}\,\text{l}^{-1}$), prior to mixing with the rest of the flow. Thereafter, treatment consisted of sedimentation, biological treatment on percolating filters with partial recirculation, settlement in humus tanks and irrigation over grass plots. Laboratory investigation showed that alum was the most effective coagulant and had the advantage that the mixed sewage had a lower pH for biological treatment. A full plant trial dosing $50\,\text{mg}\,\text{l}^{-1}$ alum as Al_2O_3 indicated that the BOD of effluent from the humus tank was reduced from $25-30\,\text{mg}\,\text{l}^{-1}$ to $15-20\,\text{mg}\,\text{l}^{-1}$. The COD was reduced from $110-135\,\text{mg}\,\text{l}^{-1}$ to $80-100\,\text{mg}\,\text{l}^{-1}$. The corresponding reduction in BOD of the grass plot effluent was from $4-6\,\text{mg}\,\text{l}^{-1}$ to $1-3\,\text{mg}\,\text{l}^{-1}$ and in COD from $50-65\,\text{mg}\,\text{l}^{-1}$ to $25-40\,\text{mg}\,\text{l}^{-1}$. Colour removal was also improved, although this proved difficult to quantify. Residual colours at various wavelengths during the alum trial were 10–20% lower following the humus tanks but were reduced by only 5–10% following the grass plots. Brief details were also given of the use of alum at other works, under the same water authority, treating dyewastes mixed with domestic sewage. The advantages claimed were that better quality effluents could be achieved or that the works could accept a higher loading and produce an effluent of similar quality.

(9) A number of studies have been reported in which the wastewater has been treated for recycling. The Textile Research Council (TRC, 1978) started an investigation into wastewater recycling by determining the quality of water required for different operations. The TRC concluded that not only could dyebaths be reused several times by making up the dye strength, but the recycling of some effluents without any treatment was feasible if they were used for less sensitive operations.

The report commented that, in many cases, the quoted quality requirements for process waters were based on current practice or current availability, rather than on actual sensitive concentrations. Although COD or TOC was not normally quoted in process waters requirements, their removal was often considered important for effluent recycle. This may be unjustified.

Reuse of dyebaths is not strictly within the scope of this report; but it is worth noting that multiple reuse of the baths did produce a stronger final effluent, but the increased strength was not proportional to the number of cycles. This approach, coupled with other water economy measures, can obviously have an impact on the required size of the effluent treatment plant.

The report went on to describe the 'Pudsey Project' in which treated sewage effluent, containing a proportion of mill effluent, was reused in a woollen mill. In this study, also reported by Harker (1980), sewage treatment consisted of primary sedimentation, partial aeration with activated sludge, settlement, biological filtration, final sedimentation, sand filtration and chlorination. The treated sewage effluent was found, in plant trials, to be satisfactory for dyeing and washing operations. This usage is gradually being extended to full-scale replacement of most of the water requirements of the mill.

The TRC study also examined the treatment, with alum, of dyebath wastewaters (used once or seven times) in order to allow their further reuse. Recycling to prepare a dyebath for a paler shade was found to be unsatisfactory. Extension of this treatment to dyewastes from five works (two already partially treated) by flocculation with $120\,\mathrm{mg\,l^{-1}}$ alum (as Al_2O_3) and addition of $4\,\mathrm{g\,l^{-1}}$ of powdered activated carbon gave mixed results. In most cases residual colour was equivalent to that in process water. However, reused water in dyeing trials was not satisfactory in every case, indicating possible interference by contaminants other than residual colour. Reuse and recycling without treatment showed the largest cost savings of £0.36–0.57 m^{-3}. Recycling of treated effluent still showed a saving of £0.21 m^{-3}, the reduction being due to the capital and operating costs of treatment.

(10) Turner (1978) describes a textile printing works in the UK which recycled effluent arising from roller washing. Aluminium and iron coagulants were investigated at various doses. The final treatment chosen consisted of coagulation with alum, lime and polyelectrolyte followed by flotation. The treated water usually had a residual pink colour whatever the colour of the untreated effluent (a common observation after alum treatment). The residual colour did not affect white untreated cloth and the water could be recycled. Discharge costs for the untreated effluent were about £0.15 m^{-3}. The charge to discharge the treated effluent was £0.037 m^{-3}. If only a proportion of the treated effluent were to be discharged as blowdown, the main part being recycled, the total costs would be much reduced.

(11) The use of coagulation with alum and sedimentation followed by ion exchange as a means of treating effluent for recycling is described by Jørgensen (1974). Coagulation was with $500\,\mathrm{mg\,l^{-1}}$ alum, the pH being adjusted to 6 with caustic soda. After 2 hours settlement the waste was still coloured and was passed through a cellulose-based cation exchanger and an anion resin. The first bed also acted as a filter. Carbon dioxide was

added prior to the cation exchanger and to the final effluent in order to maintain the pH value in the range 6–8. The alum sludge was concentrated from 2–2.5% to 15% solids by centrifuge. COD was reduced by 90–95% and the recovered water was colourless. Water was recycled up to five times and used for dyeing with no detrimental effects. The annual cost of resin replacement for treatment of $25\,m^3\,h^{-1}$ was estimated to be 0.12 Danish Kroner (DKr) per m^3. Total running costs (including generation of carbon dioxide) were about $0.6\,DKr\,m^{-3}$. This included power costs at $0.10\,DKr\,kWh^{-1}$. Inclusion of the capital costs increased the total cost to $1.06\,DKr\,m^{-3}$ (excluding sludge handling), approximately the same cost as buying municipal water. Waste treatment was therefore paid for by the saving in water costs.

(12) Another case of treatment of mixed textile waste and domestic sewage in Dalton, Georgia, USA is reported by Duchon and Painter (1978). In this town about 80% of the total waste flow was from the textile industry, mainly from the manufacture of tufted carpets. The carpets were usually dyed after stitching and dyewaste constituted a large proportion of the load. Carpets were manufactured mainly from nylon and polyester and the dyes used were predominantly disperse, acidic and cationic. In addition to the dyes, sequestrants, surfactants, scouring agents and thickeners were present. The flow to the sewage works varied between $30\,000\,m^3\,d^{-1}$ and $144\,000\,m^3\,d^{-1}$ as a result of the daily variation in the work pattern. Monthly average BOD concentrations ranged between 180 and $330\,mg\,l^{-1}$ but were increasing due to water conservation at the mills.

Treatment originally consisted of primary sedimentation, high-rate biological filtration, activated sludge and final sedimentation. Sludge was thickened by dissolved air flotation and anaerobically digested. Short textile fibres (lint) were a source of trouble at the plant. Dischargers were meant to screen the waste but sufficient fibres still passed into the sewer and through the primary sedimentation tank to cause binding of the media in the biofilter. They also wrapped around submerged equipment in the activated sludge plant. However, the main problems occurred in the digestors where accumulations had to be cleaned out every two years. At a new treatment works in the same town the problems were even worse. The plants were considered to be satisfactory until 1976 when the effluent from the second plant deteriorated more than expected under an increased load. Various possible causes were investigated, without success, including nutrient deficiency, filamentous sludge bulking and short-circuiting in the activated sludge plant. Most of the deterioration was eventually shown to be due to the overloading of the final sedimentation tank. Poor sludge-settling characteristics, due to insufficient oxygen supply to the aeration basins, also contributed.

(13) Some examples of the direct reuse of effluents are given by Porter and Sargent (1977), Goodman and Porter (1980) and Dürig et al. (1978). These authors also draw attention to the advantages of economizing by the reuse of dyebaths, while reviewing the water qualities required for different processes. The reuse of dyebaths has also been investigated by Cook and Tincher (1978), Carr and Cook (1980) and Cook et al. (1980). In the last of these papers, a case is reported in which nylon 6 and nylon 66 were dyed in baths made up to the required concentrations each time. Baths were reused for an average of thirty batches in some cases. Dye usage was reduced by 19%, other chemicals by 35%, energy costs by 57% and water and effluent disposal costs by 43%.

4.9 Future developments and trends

The main discernible trend in the manner in which textile wastes are handled is a greater emphasis on reuse and recycling. Savings of energy, chemicals, water and effluent disposal costs are the main benefits but the price is a more sophisticated effluent treatment plant. For many works, especially the smaller ones, the target will remain the minimum treatment necessary to comply with the local pollution legislation. This may take the form of a biological treatment plant or a coagulation and sedimentation plant. The choice will be partly technical and partly economic (see, for example, Clough, 1978).

The addition of activated carbon to an activated sludge plant is likely to be restricted to those cases where additional treatment is required in order to produce an effluent for recycle. The activated carbon may overcome the disadvantage of the low biodegradability of dyes, etc., by adsorbing them, but the costs are high. This cost can be justified if savings can be made in subsequent treatment, for example, by reducing the size or operating costs of a granular carbon adsorber or a resin system.

Granular activated carbon will probably remain the main form of additional treatment, especially at larger works where regeneration on site is economic. Ozonation is also finding wider application but the capital and operating costs of both these treatments are high. Alternative adsorbents based on synthetic resins are also expensive to install. Regeneration is costly and produces a residue requiring disposal. The main limitation of activated carbon is its poor adsorbance of some dyes and its inability to adsorb inorganic solutes. For these reasons, reverse osmosis may find wider application. Developments in membrane technology and the increasing market should help to make the process more competitive. However, the energy costs are comparatively high and there is still a residual effluent requiring disposal. The overall quality of the purified water is the best that can be achieved from any of the treatments described but for many applications this quality may be unnecessarily high.

A reassessment of the quality of water required for the different stages in processing textiles seems the logical step forward. In many cases the quality used is that available from the mains or works supply and this is assumed to be necessary to maintain the quality of the finished products. It has now been demonstrated that lower-quality water can be used for many processes previously thought to require mains water. Consequently, segregation of wastes can allow recycling of water with the minimum of treatment or even no treatment at all.

The environmental pressures to reduce pollution from dyewastes, especially that due to colour and organic load, will continue. If treatment has to be introduced to meet new discharge standards, the opportunity can be taken to rethink radically the ways in which the effluents are mixed and treated. Expenditure on new plant can be viewed as an opportunity to reuse and recycle, so that costs overall are minimized.

References

ADMI (1973/1974) *Dyes and the Environment*, vols 1 and 2, American Dye Manufacturers Institute.

ADMI (1978) *Textile Dyeing Wastewaters: Characterization and Treatment*, American Dye Manufacturers Institute, Rep. No. EPA 600/2-78-098, US Environmental Protection Agency, Washington, DC.

Akhmedov, K.M. and Garibov, I.M. (1966) Methods for removing sulphur dyes from wastes from the dyeing plant, *Tekh. Prog.*, 5, 41.

Alexander, F. and McKay, G. (1977) Kinetics of the removal of basic dye from effluent using silica, *Chem. Eng.*, 243–246.

APHA (1976) *Standard Methods for the Examination of Water and Wastewater*, 14th edn, American Public Health Association, American Waterworks Association and Water Pollution Control Federation, Washington, DC.

Anliker, R. (1977) Colour chemistry and the environment, *Rev. Prog. Coloration*, 8, 60–72.

Anon. (1977) Treatment of textile effluents: bleaching, dyeing, printing and finishing, *Technique de l'Eau*, No. 371, 25–34.

Anthony, A.J. (1977) Characterization of the impact of coloured wastewaters on free-flowing streams, *Proc. 32nd Purdue Ind. Waste Conf.*, Ann Arbor Science, Ann Arbor, Michigan, 288–293.

Atkins, M.H. and Lowe, J.F. (1979) *Case Studies in Pollution Control Measures in the Textile Dyeing and Finishing Industries*, Pergamon Press, Oxford.

Ayers, F.A. (Ed.) (1979) *Proc. Symp. Textile Industry Technology, Williamsburgh, Va., 1978*, Rep. No. EPA 600/2-79-104, US Environmental Protection Agency, Washington, DC.

Blank, H.U., Keller, W. and Kuhne, G. (1976) Process for Decolorising Effluents, Brit. Pat. 1 490 691.

Bettens, L. (1979) Powdered activated carbon in an activated sludge unit, *Effluent Water Treat. J.*, March, 129–135.

Boudreau, Dubeau, Lemieux Inc. (1981) *Textile Industry Wastewater Treatment by Air Flotation*, Rep. No. EPS-4-WP-79-6E, Environment Canada, Ottawa.

Brandon, C.A. (1980) Closed-cycle textile dyeing: full scale hyperfiltration demonstration, *Nat. Water Supply Improvement J.*, **7**, 39–49.

Brown, D., Hitz, H.R. and Schäfer, L. (1981) The assessment of the possible inhibitory effect of dyestuffs on aerobic waste-water bacteria: experience with a screening test, *Chemosphere*, **10**, 245–261.

Carr, W.W. and Cook, F.L. (1980) Savings in dyebath reuse depend on variations in impurity concentrations, *Text. Chem. Color.*, **12**, 106–110.

Cheremisinoff, P.N. and Ellerbusch, F. (1978) *Carbon Adsorption Handbook*, Ann Arbor Science, Ann Arbor, Michigan.

Ciba Geigy AG (1975) Process for Purifying Aqueous Industrial Effluents, Brit. Pat. 1 449 387.

Clough, G.F.G. (1978) Chemical versus biological effluent treatment, in *Chemistry of Effluent Treatment*, Kakabadse, G. (Ed.), Applied Science, London, 27–37.

Cohen, H. (1975) The use of ultrafiltration membranes in the treatment of textile dyehouse waste, *2nd Nat. Conf. on Complete Water Reuse, Chicago*, Amer. Inst. Chem. Eng. and US Environmental Protection Agency, Washington, DC, 992–995.

Cook, F.L. and Tincher, W.C. (1978) Dyebath reuse in batch dyeing, *Text. Chem. Color.*, **10**, 21–25.

Cook, F.L., Tincher, W.C., Carr, W.W., Olson, L.H. and Averette, M. (1980) Plant trials on dyebath reuse show savings in energy, water, dyes and chemicals, *Text. Chem. Color.*, **12**, 15–24.

Crowe, T., O'Melia, C.R. and Little, L. (1977) The coagulation of disperse dyes, *Proc. 32nd Purdue Ind. Waste Conf.*, Ann Arbor Science, Ann Arbor, Michigan, 655–662.

DeJohn, P.B. (1976) Factors to consider when treating dyewastes with granular activated carbon, *Proc. 31st Purdue Ind. Waste Conf.*, Ann Arbor Science, Ann Arbor, Michigan, 375–384.

DiGiano, F.A. and Natter, A.S. (1977) Disperse dye–carrier interactions on activated carbon, *J. Water Pollut. Cont. Fed.*, **49**, 235–244.

Dohányos, M., Maděra, V. and Sedláček, M. (1978) Removal o' organic dyes by activated sludge, *Prog. Water Technol.*, **10**, 559–575.

Duchon, K. and Painter, M. (1978) Experience with combined tufted textile–municipal wastewater treatment facilities in Dalton, Georgia, *Proc. 33rd Purdue Ind. Waste Conf.*, Ann Arbor Science, Ann Arbor, Michigan, 744–757.

Dürig, G., Hausmann, J.P. and O'Hare, B.J. (1978) A review of the possibilities for recycling aqueous dyehouse effluent, *J. Soc. Dyers Colourists*, **94**, 331–338.

Éndyus'kin, P.N., Filippov, V.M. and Dyumaev, K.M. (1979) Use of active chlorine for decolorization of wastewaters from production of organic dyes, *J. Appl. Chem. (USSR)*, **52**, 783–787.

EPA (1974) *In-plant Control of Pollution: Upgrading Textile Operations to Reduce Pollution*,

Rep. No. EPA-625/3-74-004, US Environmental Protection Agency, Washington, DC.

EPA (1979) Textile mill point source category effluent limitation guidelines, pretreatment standards and new source performance standards, Fed. Reg. 44, 62 204-62 241, US Environmental Protection Agency, Washington, DC.

EPA (1980a) *Treatability Manual:* Vol. 2 *Industrial Descriptions*, Rep. No. EPA-600/8-80-042b, US Environmental Protection Agency, Washington, DC.

EPA (1980b) *Treatability Manual:* Vol. 3 *Technologies for Control/Removal of Pollutants*, Rep. No. EPA-600/8-80-042c, US Environmental Protection Agency, Washington, DC.

Erndt, E. and Kurbiel, J. (1980) The application of ozone in removal of refractory substances from textile wastewater, *Environ. Prot. Eng.*, **6**, 19–35 (in Polish).

Fisons Ltd (1976) Process for Treating Aqueous Effluent Containing Dyes, Brit. Pat. 1 544 430.

Forster, C.F. (1977) Bio-oxidation, in *Treatment of Industrial Wastes*, Callely, A.G., Forster, C.F. and Stafford, D.A. (Eds), Hodder and Stoughton, London, 65–87.

Freytag, R. and Diemunsch, J. (1979) Problems of effluent treatment in the cotton finishing industry, *Melliand Textilber.* (English edn), **60**, 598–602.

Friedman, M., Diamond, M.J. and MacGregor, J.T. (1980) Mutagenicity of textile dyes, *Environ. Sci. Technol.*, **14**, 1145–1146.

Fung, D.Y.C. and Miller, R.D. (1973) Effects of dyes on bacterial growth, *Appl. Microbiol.*, **25**, 793–799.

Gardiner, D.K. and Borne, B.J. (1978) Textile waste waters: treatment and environmental effects, *J. Soc. Dyers Colourists*, **94**, 339–348.

Goodman, G.A. and Porter, J.J. (1980) Water quality requirements for reuse in textile dyeing processes, *Am. Dyest. Rep.*, October, 33–39 and 46.

Harker, R.P. (1980) Recycling sewage water for scouring and dyeing, *Am. Dyest. Rep.*, January, 28–37 and 46.

Hitz, H.R., Huber, W. and Reed, R.H. (1978) The adsorption of dyes on activated sludge, *J. Soc. Dyers Colourists*, **94**, 71–76.

Hobson, R.J. (1977) Effluent treatment by oxyphotolysis, *Pollution Monitor*, August/September, 12.

Hocke, B. (1978) Methods for the reduction of COD in textile effluents, *Textilveredlung*, **13**, 232–236.

Jackson, G.E. (1980a and 1980b) Granular media filtration in water and wastewater treatment, Parts 1 and 2, *CRC Crit. Revs. Environ. Control*, **10**, 339–373 and **11**, 1–36.

Jørgensen, S.V. (1974) Recirculation of waste water from the textile industry is possible when a combined precipitation and ion exchange treatment is used, *Vatten*, **4**, 364–369.

Judkins, J.F. and Hornsby, J.S. (1978) Color removal from textile dyewaste using magnesium carbonate, *J. Water Pollut. Cont. Fed.*, **50**, 2446–2456.

Kace, J.S. and Linford, H.B. (1975) Reduced cost flocculation of a textile dyeing wastewater, *J. Water Pollut. Cont. Fed.*, **47**, 1971–1977.

Kitamura, K., Takabayashi, F., Shibata, F., Watabe, K. and Azumi, T. (1979) Halohydrocarbon Treatment of a Glycol and Waste Dye Liquor Followed by Water Extraction of the Glycol, US Pat. 4 165 217.
Koziorowski, B. and Kucharski, J. (1972) *Industrial Waste Disposal*, Pergamon, Oxford and Wydawnictwa Naukowo-Technicze, Warsaw.
Leslie, M.E. (1975) Water treatment with peat moss solves textile dye house pollution problems, *Can. Textile J.*, **92**, 67–70.
Little, A.H. (1967) Treatment of textile waste liquors, *J. Soc. Dyers Colourists*, **83**, 268–273.
Little, A.H. (1975) *Water Supplies and the Treatment and Disposal of Effluents*, Textile Institute Monograph No. 2, Textile Institute, Manchester.
Little, L.W., Lamb, J.C., Chillingworth, M.A. and Durkin, W.B. (1974) Acute toxicity of selected commercial dyes to the fathead minnow and evaluation of biological treatment for reduction of toxicity, *Proc. 29th Purdue Ind. Waste Conf.*, Ann Arbor Science, Ann Arbor, Michigan, 524–534.
Maggiolo, A. and Sayles, J.H. (1977) *Automatic Exchange Resin Pilot Plant for Removal of Textile Dye Wastes, Bennett College, Greensboro, N. Carolina*, EPA Rep. No. 600/2-77-136, US Environmental Protection Agency, Washington, DC.
Mahloch, J.L., Shindala, A., McGriff, E.C. and Barnett, W.A. (1974) Treatability studies and design considerations for a dyeing operation, *Proc. 29th Purdue Ind. Waste Conf.*, Ann Arbor Science, Ann Arbor, Michigan, 44–50.
McKay, G. (1979a) Waste colour removal from textile effluents, *Am. Dyest. Rep.*, April, 29–36 and 47.
McKay, G. (1979b) Coloured effluents — environmental and legal aspects, *Water Waste Treatment*, March, 37–41.
McKay, G., Otterburn, M.S. and Sweeney, A.G. (1978) The removal of colour from effluent using various adsorbents: some preliminary economic considerations, *J. Soc. Dyers Colourists*, **94**, 357–360.
McKay, G., Otterburn, M.S. and Sweeney, A.G. (1980a) The removal of colour from effluents using various adsorbents: III Silica: rate process, *Water Res.*, **14**, 15–20.
McKay, G., Otterburn, M.S. and Sweeney, A.G. (1980b) The removal of colour from effluents using various adsorbents: IV Silica: equilibrium and column studies, *Water Res.*, **14**, 21–27.
Michaels, G.B. and White, E.M. (1978) *Microbial Degradation of Dyewastes in Aqueous Effluent, Environmental Resources Center, Institute of Technology, Atlanta, Ga.*, NTIS No. PB-293 219, Springfield, Virginia.
Mitchell, M., Ernst, W.R., Rasmussen, E.T., Bagherzadeh, P. and Lightsey, G.R. (1978) Adsorption of textile dyes by activated carbon produced from agricultural, municipal and industrial wastes, *Bull. Environ. Contam. Toxicol.*, **19**, 307–311.
Murakami, K. and Annaka, T. (1978) Removal of color from municipal sewage containing textile wastewater, *Proc. 6th US/Japan Conf. on Sewage Treatment Technology*, Cincinnati, NTIS No. PB-80-177595, Springfield, Virginia, 199–224.
Mutch, N. (1946) Dried alumina, *J. Pharm. Pharmacog.*, **19**, 490–519.
Nebel, C. and Stuber, L.M. (1976) Ozone decoloration of secondary dye-laden

effluents, *Proc. 2nd Internat. Symp. on Ozone Technology, Montreal, 1975*, Int. Ozone Inst., 336–358.

Nemerow, N.L. (1978) *Industrial Water Pollution: Origins, Characteristics and Treatment*, Addison-Wesley, Reading, Mass.

Netzer, A. and Beszedits, S. (1975) Colour and heavy metal removal from dyebath effluents by lime precipitation, *Proc. 10th Can. Symp. Water Pollut. Res.*, University of Toronto, 151–163.

OECD (1980) *Water Management in Industrialised River Basins*, Organisation for Economic Cooperation and Development, Paris.

OECD (1981) *Emission Control Costs in the Textile Industry*, Organisation for Economic Cooperation and Development, Paris.

Parish, G.H. (1977) Textile and tannery wastes, in *Treatment of Industrial Effluents*, Callely, A.G., Forster, C.F. and Stafford, D.A. (Eds), Hodder and Stoughton, London, 229–244.

Poots, V.J.P., McKay, G. and Healy, J.J. (1976a) The removal of acid dye from effluents using natural adsorbents: I Peat, *Wat. Res.*, **10**, 1061–1066.

Poots, V.J.P., McKay, G. and Healy, J.J. (1976b) The removal of acid dye from effluent using natural adsorbents: II Wood, *Wat. Res.*, **10**, 1067–1070.

Poots, V.J.P., McKay, G. and Healy, J.J. (1978) Removal of basic dyes from effluent using wood as an adsorbent, *J. Water Pollut. Cont. Fed.*, **50**, 926–935.

Porter, J.J. and Sargent, T.N. (1977) Waste treatment versus waste recovery, *Text. Chem. Color.*, **9**, 269–273.

Porter, J.J. and Snider, E.H. (1974) Thirty day biodegradability of textile chemicals and dyes, *Rep. Nat. Tech. Conf. Amer. Assoc. Textile Chemists and Colorists*, N. Carolina, 427–436.

Porter, J.J. and Snider, E.H. (1976) Long-term biodegradability of textile chemicals, *J. Water Pollut. Cont. Fed.*, **48**, 2198–2209.

Rawlings, G.D. and Samfield, M. (1979) Textile plant wastewater toxicity, *Environ. Sci. Technol.*, **13**, 160–164.

Rinker, T.L. and Sargent, T.N. (1974) Activated sludge and alum coagulation treatment of textile wastewaters, *Proc. 29th Purdue Ind. Waste Conf.*, Ann Arbor Science, Ann Arbor, Michigan, 456–471.

Rock, S.L. and Stevens, B.W. (1975) Polymeric adsorption–ion exchange process for decolorizing dye waste streams, *Text. Chem. Colour*, **7**, 169–171.

Rohrer, E. (1977) Catalytically assisted oxidation process, *Chem. Ind.*, 816–821.

Schaffer, R.B. (1978) Water pollution control in the textile industry, in *Proc. Symp. Textile Industry Technology* (Ayers, F.A., Ed.), *loc. cit.*

Schwägler, U. and Stotz, G. (1980) Influence of microorganisms on the adsorptive clarification of waste water from textile mills, *Melliand Textilber.* (English edn), **61**, 599–603.

Sethuraman, V.V. and Raymahashay, B.C. (1975) Color removal by clays, *Environ. Sci. Technol.*, **9**, 1139–1140.

Shelley, M.L., Randall, C.W. and King, P.H. (1976) Evaluation of chemical–biological and chemical–physical treatment for textile dyeing and finishing waste, *J. Water Pollut. Cont. Fed.*, **48**, 753–761.

Shirazi, H., Maeda, Y., Mirzadeh, A., Djadali, M. and Fazeli, A. (1977) Treatability of dyeing wastewater by a combined chemical coagulation and activated sludge process, *J. Ferment. Technol.*, **55**, 249–257.

Sidwick, J.M. and Barnard, R. (1981) Treatment before discharge, *Chem. Ind.*, 277–285.

Singer, P.C. (1976) Dye study focuses on treatment, *ESE Notes*, Dept. Environmental Science and Engineering, School of Public Health, University of North Carolina at Chapel Hill, **12**, 1–2 and 8.

Snider, E.H. and Porter, J.J. (1974) Ozone treatment of dye waste, *J. Water Pollut. Cont. Fed.*, **45**, 886–894.

TRC (1978) *Water Quality Requirements and Wastewater Recycling in the UK Textile Industry*, Textile Research Council (UK), Nottingham.

Thornton, H.A. and Moore, J.R. (1951) Adsorbents in waste water treatment — Dye adsorption and recovery studies, *Sewage Ind. Wastes*, **23**, 497–504.

Trotman, E.R. (1975) *Dyeing and Chemical Technology of Textile Fibres*, 5th edn, Charles Griffin, London.

Turner, M.T. (1978) Wastewater treatment and reuse within the textile industry, *Water Serv.*, **82**, 527–528.

Voorn, G. (1976) Purification with activated carbon of textile industry wastewater, *Science Industry*, **8**, 7–8.

Waters, B.D. (1979) Treatment of dyewastes, *Water Pollut. Control*, **78**, 12–24.

Weber, W.J. (1972) *Physicochemical Processes for Water Quality Control*, Wiley Interscience, New York.